We have to remember that what we observe
is not nature in itself,
but nature exposed
to our method of questioning.

- Werner Heisenberg

Dedications:

LP: To Pap, who taught me how to listen
& to Gram, who should have had the medicine.

SE: To my Mother & my Grandmother
whose pain I would have loved to relieve

Chapter 25:
The CROWN

Vol. 1: how to shoot an elephant

writing & images by Lex Pelger
illustration & collage by Sumacc Esquives

a standalone installment from the ongoing series
'Anandamide, or; the Cannabinoid'
(based on the 135 chapters of 'Moby-Dick, or; the Whale')

by No Nonsense Books

Many thanks to the editors & proofreaders:
Zoe Platek, Patrick Riley,
Kynthia Brunette & Scarlet Palmer

A thank you to Douglas Greene
for helping me keep my fakts straight. RIP

Publishing History:
1.0: March 2017, NYC
1.1: April 2024, Bordeaux
(updated the Turtles, added more image credits
& removed the Tobacco/Acetylcholine/Neural city
metaphor to move it forward)

For more cannabinoid science, see
Lex's newsletter 'Cannabinoids & the People'

For more from this graphic novel series,
see lexpelger.com

See Sumacc's interdisciplinary art
on IG @wayrasumaccart
or raquelesquives.com

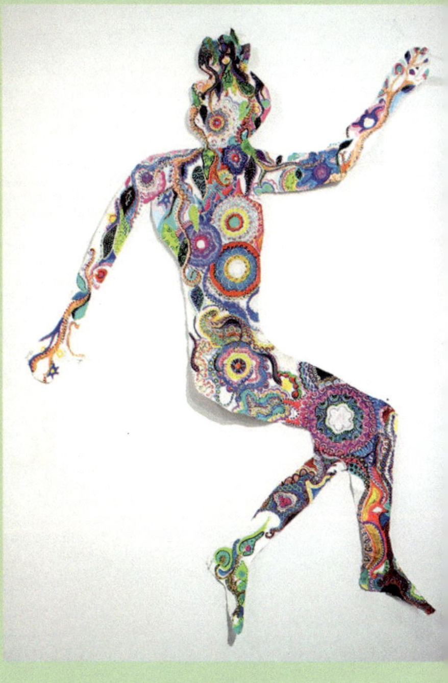

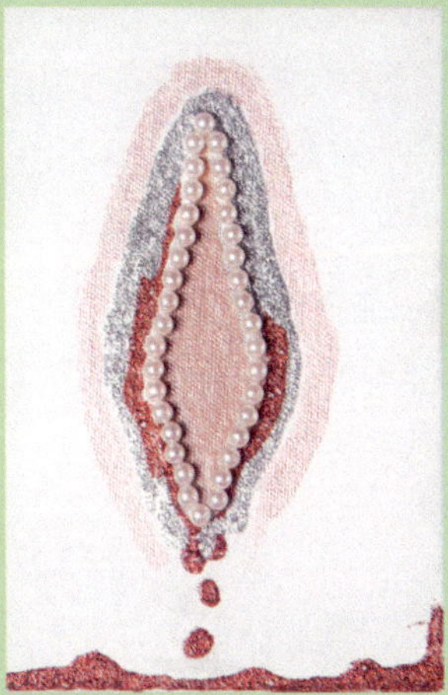

1963: THE YEAR OF THE SHIFT

A few months before the end of JFK's Camelot,

a progress report emerges from the President's ADVISORY COMMISSION ON NARCOTIC AND DRUG ABUSE

FOR RELEASE TO P.M., APRIL 4, 1963

April 3, 1963

Office of the White House Press Secretary

THE WHITE HOUSE

FOLLOWING IS AN EXCHANGE OF LETTERS

In response, John F. Kennedy personally thanks Judge E. Barrett Prettyman for the recent update on the efforts of his committee to study the United States drug problem.

Dear Judge Prettyman:

I have received your interim report and wish to thank you and the other members of the President's Advisory Commission on Narcotic and Drug Abuse for the obvious time and effort that have gone into preparing the report. This report deserves to be read, discussed and considered by those interested in our nation's drug and narcotic problems -- it begins to open the door to understanding about users of narcotics and abusers of the law and suggests forcefully the dangers which confront the American people in this difficult area.

The recommendations put forward by the Commission will be studied carefully by those departments and agencies of the Federal government charged with the health and legal protection of our citizens. Special attention should be given to those recommendations which can be translated into action programs designed to provide practical, workable solutions to the many aspects of narcotic and drug abuse. You may be sure that the various agencies of the Federal government will cooperate fully in giving needed assistance in the preparation of your final report -- a report which I look forward to with great interest.

Sincerely,

s/ John F. Kennedy

Just six months previously,
as January dawned on that momentous year of 1963,
JFK had issued his executive order to form a blue-ribbon commission
to grapple with this perennial Amerikan problem.

= ENDNOTE

PRESIDENT'S ADVISORY COMMISSION

ON NARCOTIC AND DRUG ABUSE

THE UPDATE SEEN BY KENNEDY
MATCHED GREATLY TO THE FINAL PRODUCT
& THUS IT'S APPROPRIATE THAT THE
INTERIM REPORT
OFFICIALLY COMPLETED ON
APRIL 1, 1963
BECAUSE ITS SUGGESTIONS ON THE QUESTION
OF AMERIKAN PSYCHOACTIVES & THEIR ABUSE
STRIKES ONE AS SO FORWARD THINKING
AMIDST THE HEARTLESS DRUG POLICIES
STILL SUPREME AFTER THE SEVERE 1950S
THAT IT MIGHT WELL HAVE BEEN A JOKE
OF APRIL FOOLS.

Unfortunately, the release of Prettyman's final report in January 1964 missed the 35th Executive by just three months...

& instead went to the desk of President Lyndon Baines Johnson, known sometimes affectionately as the Hammer.

The report emphasized the committee's many questions – but to those inquiries, they found more disputes than answers.

For instance, at that time the number of opiate addicts was declining across the nation while the use of non–narcotic drugs was up.

Why?

The commission could only state:
"as for the causes...
there is pitifully little precise information."

Interviewing experts, the committee learned that the incidence of drug abuse usually appeared in the late teens, rose rapidly through the twenties & then declined after the age of forty.
They found the price of heroin had jumped since WWII
while the purity had crashed.

They read reports from the mayors of big cities like NYC & LA who complained of spikes in petty crime that financed private drug habits.
They heard tell of the recent mafia busts that proved correct the ancient allegations of the Federal Bureau of Narcotics: the Cosa Nostra had finally entered the high–level drug smuggling game.
The Mafia's traditional code of honor against working the 'vice rackets' could not resist succumbing to that alluring glitter of gold to be harvested from yet another Amerikan Prohibition.

The commission stressed the need for accurate knowledge

to "combat the misinformation so prevalent."

As for the users themselves, no one even knew the number of those "habituated to drugs" & no one had data on the recidivism rates of the treatment programs.

The committee members emphasized "it should be remembered that these drugs, if properly used, are medically valuable".

They wanted to avoid laws that penalized the users, who they referred to as "victims, a term that today might be seen as troublesome to some harm reductionists, but in the drug abuse dark ages of the '50s & early '60s, these statements seem luminous with compassion.

For possession of drugs, Prettyman's commission advocated for the abolishment of mandatory minimum sentences, a new legal regime that had begun clogging the courts & prisons. These laws also neutered the judgment of judges wishing to assign more nuanced paroles & penalties for drug users, depending on the unique situation of the person before their bench.

They even went so far as to float the idea of federal courts offering drug addicts "treatment & care in lieu of prosecution for the offense" at facilities like the Narcotic Farm in Lexington seen here (though this forced them to ask the Department of Justice for a report on the constitutionality of the government mandating the civil commitment of drug users).

And finally, they brought to light the obvious question...

What to do about the rising tide of new drugs & the old formulations now sold under new guises?

The uppers masquerading as diet pills, the barbiturates downing the nation,

& in that banner year of 1963, the launch of a new fixture in Amerikan drug-guzzling taking off like a psychoactive Apollo mission: Valium.

Valium spent 13 years at #1
– the top–selling legal psychoactive in the United States –
& powered its manufacturer Roche into a pharmaceutical behemoth.
In 1978, when the US population was ~222 million,
Roche sold 2.2 billion Valium tablets,
100 pills per person per year,
or, one diazepam every three days for every human in the country.
Huxley saw it coming. None of Orwell's coercive techniques needed.
Aldous knew we'd be queuing up for our soma.

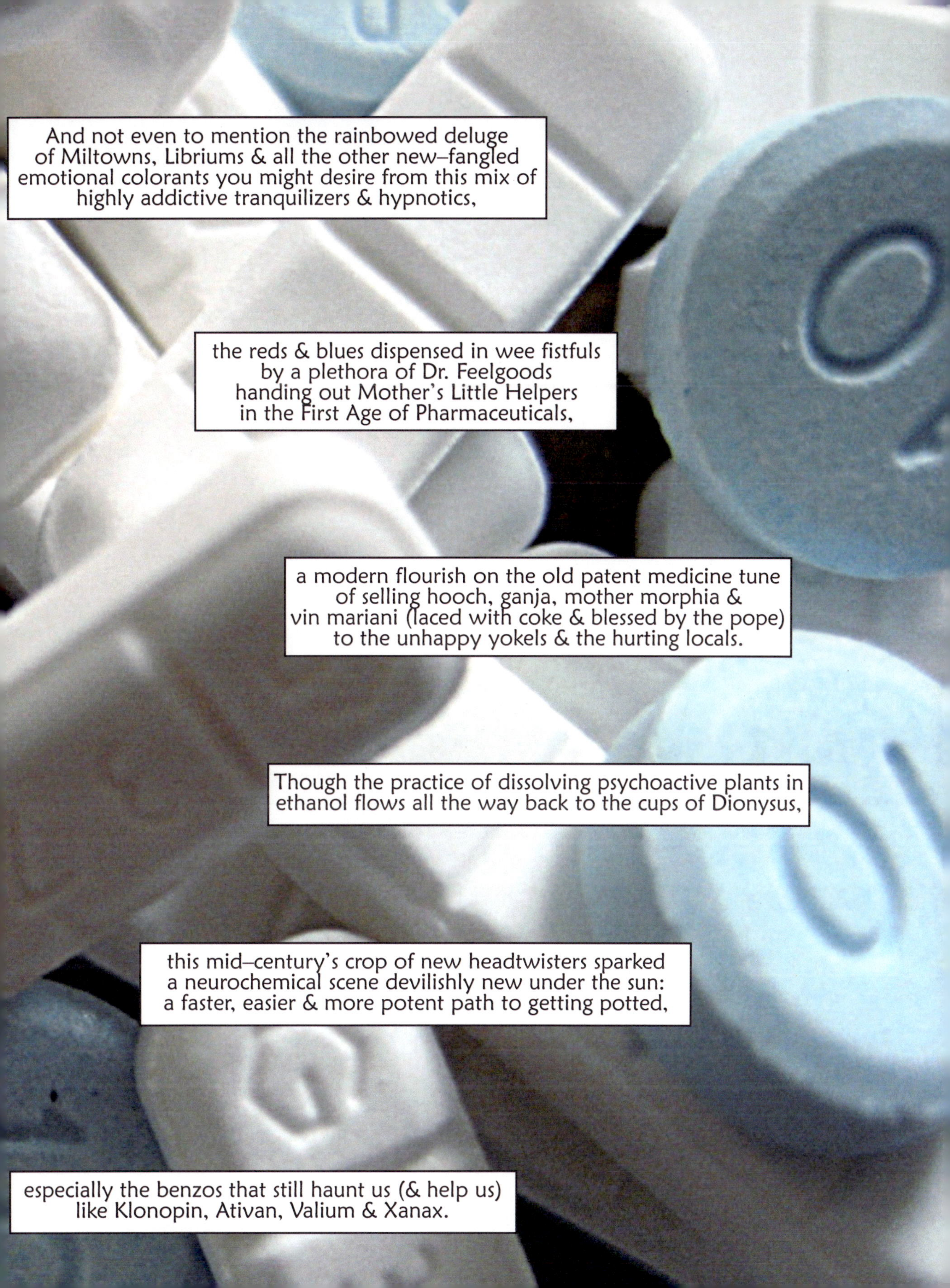

And not even to mention the rainbowed deluge of Miltowns, Libriums & all the other new-fangled emotional colorants you might desire from this mix of highly addictive tranquilizers & hypnotics,

the reds & blues dispensed in wee fistfuls by a plethora of Dr. Feelgoods handing out Mother's Little Helpers in the First Age of Pharmaceuticals,

a modern flourish on the old patent medicine tune of selling hooch, ganja, mother morphia & vin mariani (laced with coke & blessed by the pope) to the unhappy yokels & the hurting locals.

Though the practice of dissolving psychoactive plants in ethanol flows all the way back to the cups of Dionysus,

this mid-century's crop of new headtwisters sparked a neurochemical scene devilishly new under the sun: a faster, easier & more potent path to getting potted,

especially the benzos that still haunt us (& help us) like Klonopin, Ativan, Valium & Xanax.

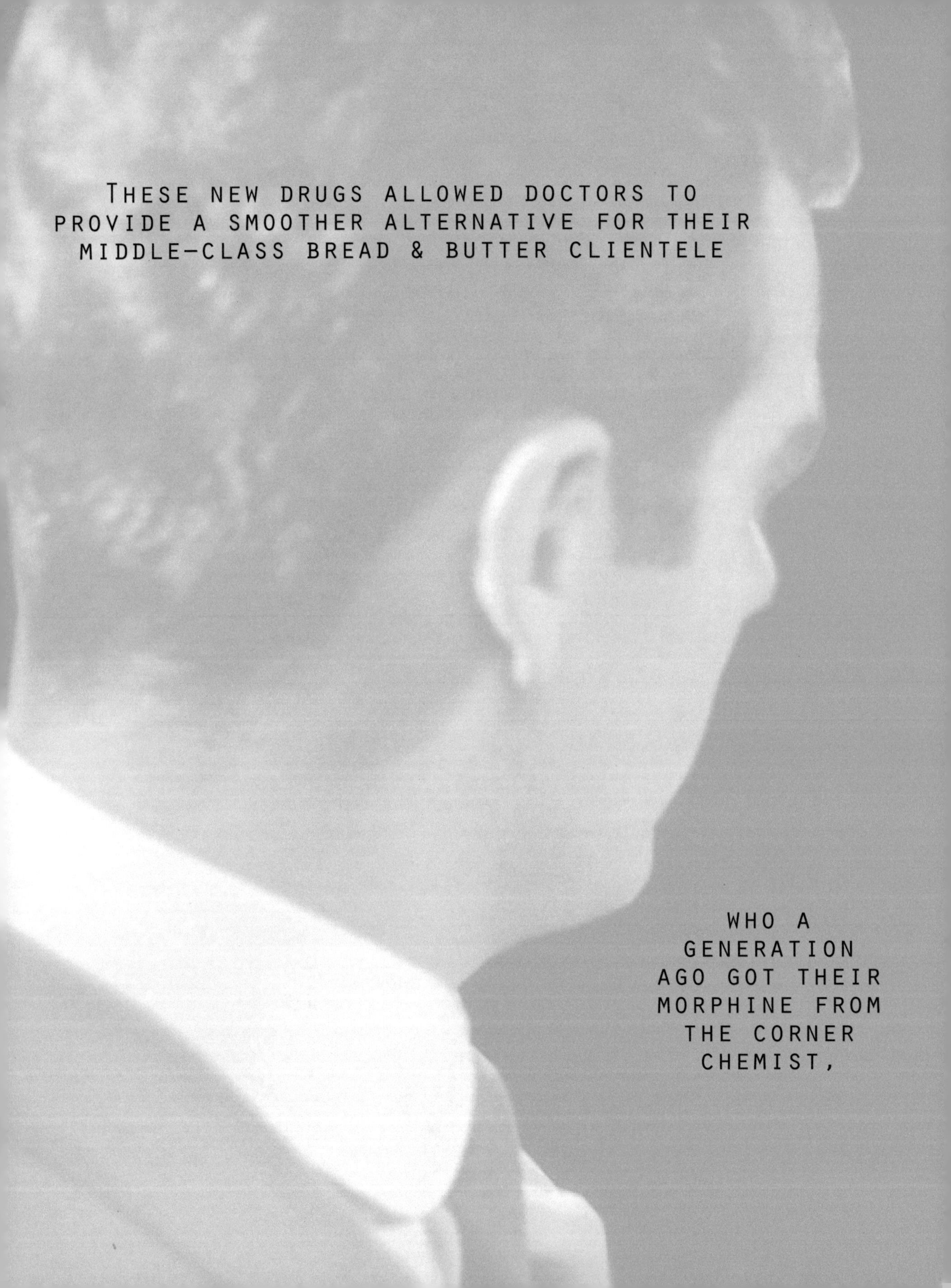

THESE NEW DRUGS ALLOWED DOCTORS TO PROVIDE A SMOOTHER ALTERNATIVE FOR THEIR MIDDLE-CLASS BREAD & BUTTER CLIENTELE

WHO A GENERATION AGO GOT THEIR MORPHINE FROM THE CORNER CHEMIST,

A SITUATION THAT SET SOME,
LIKE THE MOTHER OF PLAYWRIGHT
EUGENE O'NEILL...

DOWN A LONG DAY'S JOURNEY INTO NIGHT.

While that old addictive drama of Middle Amerika repeated itself, this time with new & respectable brand names, the face of opiate dependence in the public mind gradually transformed from the well–featured ladies of the parlor rooms...

Harry Anslinger used his helm at the Federal Bureau of Narcotics to keep his agents on a pogrom against any physicians seen as too liberal with the opiate scripts.

This trepidation about dispensing painkillers led to many patients suffering from legitimate sources of pain like kidney stones.

Because that terrible condition was a classic untestable diagnosis for those looking to score, many a young man was forced to endure the agony without any medication to alleviate their very real & severe pain (an old war veteran who'd been shot 3 times told me that he'd rather get shot again than have kidney stones).

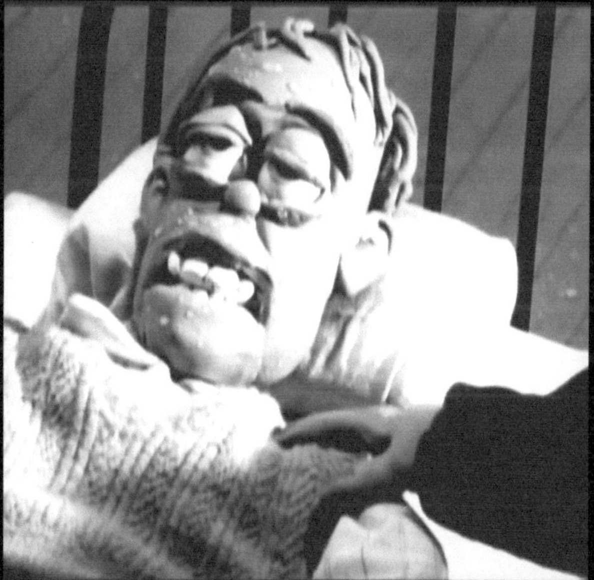

This phenomenon of an enforced shortage only repeats more widely today with certain classes & certain continents entirely bereft of opiate painkillers.

When the Harrison Act of 1914 first kindled federal drug control, the law's wording allowed a physician to dispense narcotics in "professional practice only" & for "legitimate medical purposes".

These two wholly undefined phrases, fuzzy with disagreements between medical doctors, federal narcotic law enforcers & health researchers led to the narcotic problem then being just as unanswered as today.

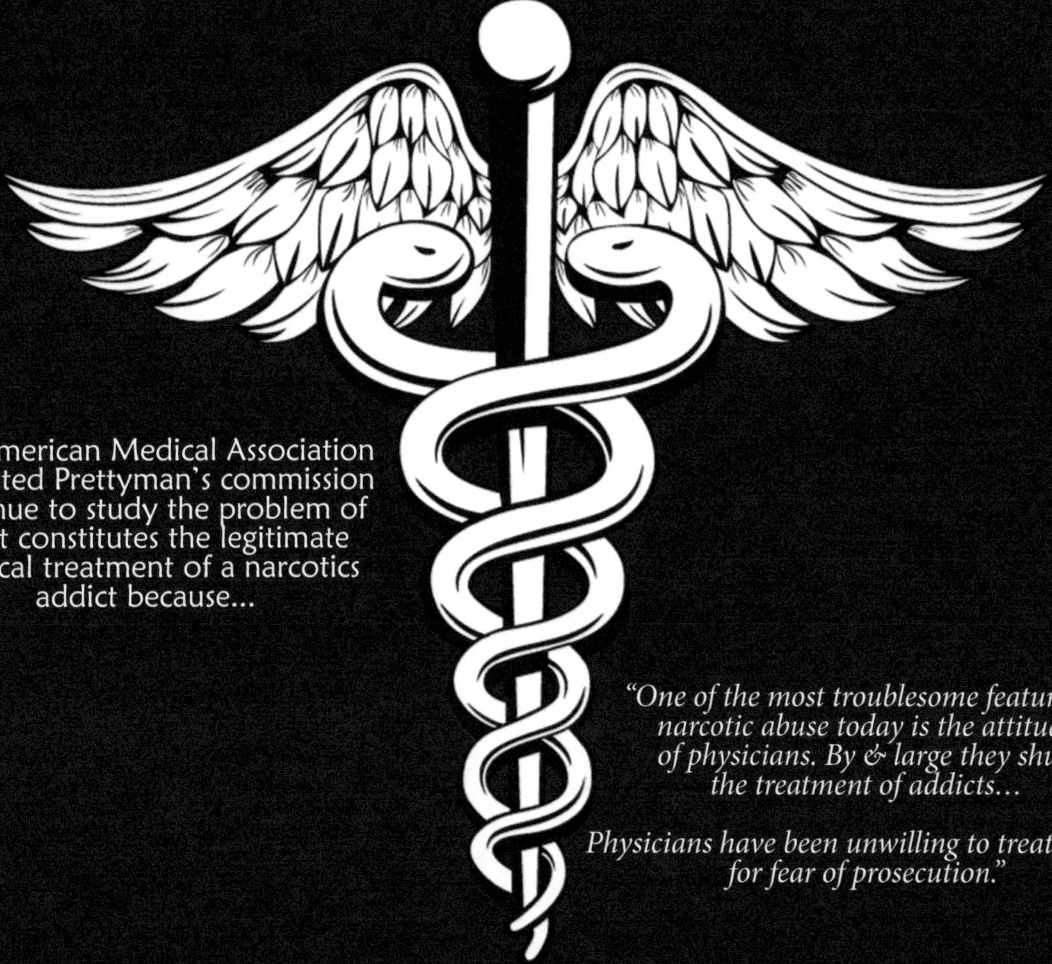

The American Medical Association requested Prettyman's commission continue to study the problem of what constitutes the legitimate medical treatment of a narcotics addict because...

"One of the most troublesome features of narcotic abuse today is the attitude of physicians. By & large they shun the treatment of addicts...

Physicians have been unwilling to treat addicts for fear of prosecution."

Just previously in 1962, a staunch wing of the AMA went on record to correct any possible impression they might favor the establishment of "community ambulatory clinics", outpatient treatment centers dispensing maintenance doses of opiates to patients (this treatment option being the current gold standard model of harm reduction in Switzerland & a few other humanely minded countries).

Heroin use took up a great deal of column inches in the press –
but the actual users only consisted of quite a small number of people.

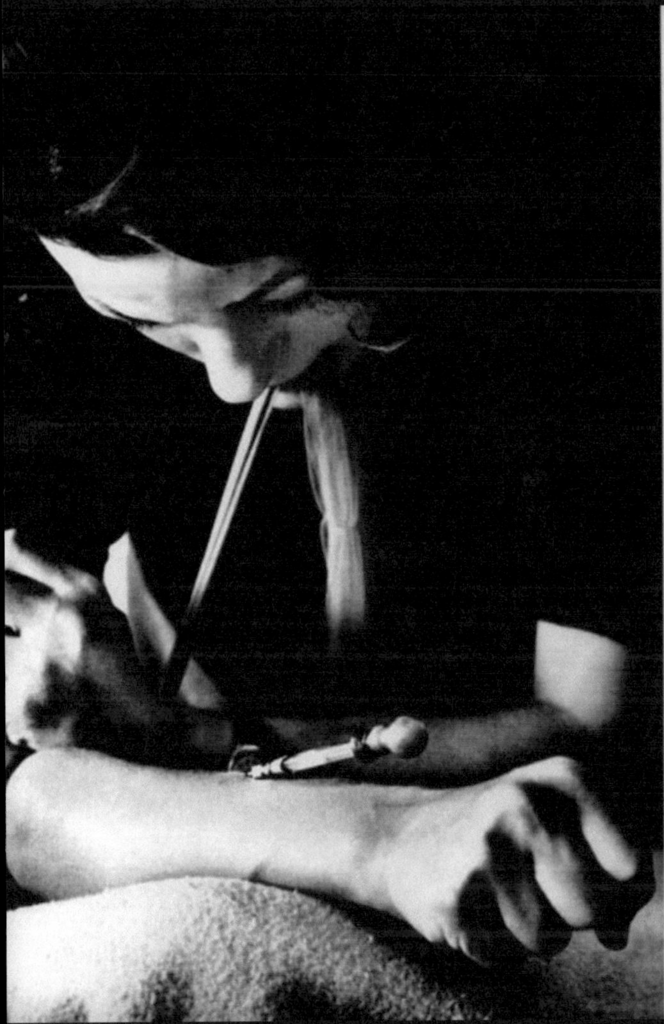

PART I

John and Karen, Two Lives Lost to Heroin

Four times a day, this is John and Karen, holed up with heroin. Faces desperate and intent, teeth pulling tight the tourniquet, grimy fingers squeezing fluid into the bloodstream, and then—peace. This is everything they live for—this is what heroin is all about. With this shot their problems vanish, and the world they cannot handle fades to leave them in solitary bliss.

No "square"—the addict's word for anyone who does not use drugs—can imagine the strength of heroin's hold. The addict will beg for it, walk miles for it, wait hours for it, con for it, stay up days and nights on end to pursue it, steal from those he has loved for it, risk death for it. Heroin, more than any other drug, leads the greatest number of addicts to squalor and desperation.

The frightening power heroin holds over John and Karen is the subject of this essay and the article that follows. To outsiders, they sometimes seem to stroll (*preceding pages*), but always they are driven by the drug—he to thievery, she to prostitution, and both to "pushing" heroin to pay for their own supply. The drug urges them, as the story shows, to murky streets and ill-lit corners, through shabby rooms and in and out of hospitals and jails. It is their jealous lover, and their wrathful god.

Every day heroin wins a few new converts to its ranks, and now there are more addicts in America than authorities can successfully count. The Federal Narcotics Bureau estimates that the U.S. has 60,000 heroin addicts, but other less official counts climb into the hundreds of thousands. Half of the country's addicts live in New York City, and almost all the others are in the slums of Detroit, Chicago and Los Angeles. Half are Negro. Only 20% are women. They commit an enormous number of crimes—more than 15% of New York City's burglaries (but less than 2% of its felonious assaults). Few are violent. Contrary to popular belief, it is not heroin that may lead to violence, but the excessive use of other drugs: amphetamines, barbiturates, cocaine, Doriden, marijuana.

The heroin addict is a very busy man. For those who would separate him from his heroin he has no use and no time. When he awakes in the morning he reaches instantly for his "works"—eye-dropper, needle ("spike," he calls it), and bottle top ("cooker"). He dissolves heroin in water in the cooker and injects the mixture. This is his "wake-up," a morning shot to hold off the anxiety and sickness of withdrawal and get him "straight" enough to start the day. If his habit is costing him $20 a day, and that is not a large habit by any means, he must now start out to steal at least $100 worth of goods, knowing that a fence will give him only one-fifth the true value of his loot. When he has stolen something, he must haggle with his fence over the price. The argument seems interminable to him, for it has now been hours since his wake-up and he is getting nervous again, his eyes are watering and he is beginning to feel like a man coming down with a bad case of flu.

Finally he gets the money and begins his search for a "connection"—someone with heroin for sale. Not just any connection, but a connection who deals good-quality stuff—"dynamite," not "garbage." Once the addict has bought his fix (has "copped" or "scored") he is faced with the risky business of getting it to his cooker and into his arm without getting caught and "busted" (arrested). When he has finally injected the heroin—he calls it "shooting up," "taking off," "getting off"—he may or may not go on a "nod,"—his eyelids heavy, his mind wandering pleasantly—depending on how much heroin his body has become accustomed to and how much actual heroin was in the powder he injected.

He hopes that the shot will be at least strong enough to make him straight for a few hours. He can judge immediately the quality of the shot. If it is strong enough, he calms down, the flu feeling leaves, and he instantly begins looking for money for the next shot.

What haunts the addict are anxieties, which only heroin can relieve. In the shaky families and oppressive environment of big-city slums, anxieties pile up fast—and it is in the teeming slums that heroin is handy. From friend to friend the drug spreads inexorably among the emotionally weak and unstable.

John and Karen have much in common with other big-city junkies. Karen is 26, John 24. Both had broken the law before they started on heroin—she as a prostitute in the Midwest, he as a thief in New York. Karen is the first in her family to use illegal drugs; but John has two addicted brothers, and a third died of an overdose.

Both John and Karen have used many drugs, but they prefer heroin to all the rest just as a gourmet prefers wine to beer. Both have been to jail (he 10 times, she twice) and to hospitals (he 4 times, she twice)—and have emerged each time to start their habits fresh.

John and Karen have been together—sleeping wherever they can find a place to lie down—for three years. They use the same last name, but never got around to formal marriage ("We did get a blood test once," says Karen). Karen's earnings as a prostitute also support John's habit, and he occasionally contributes a little money by breaking into parked cabs, in which drivers may have left coin changers.

Both John and Karen are at times all but overcome by revulsion for their habit and for the horrifying, unseen world it forces them into. "We are animals," says Karen. "We are all animals in a world no one knows."

A stocking wrapped around her arm to make the vein stand out, Karen (left) waits for blood to start backing into the eyedropper—a sign the needle is in the vein. To get more pressure, she has replaced the dropper's small bulb with a nipple from a baby pacifier. John (right) also shoots directly into a vein (a practice known as "mainlining") but he does it farther up his arm.

69

Booze & tobacco continued to dominate Amerikan drug consumption
while the pharmaceutical pipelines vastly changed the contours of consciousness
by pumping out the newest neurochemical twists in floods sculpting the landscape.

BUT ALL THIS TIME,
OUR HARDY LITTLE HEMP PLANT
LURKED IN THE WINGS,
READY TO PHOENIX BACK
INTO THE HEADS
& THE HEADLINES.

HEMP

for Michael Aldritch

The generation prior, the Beats had continued the tradition of marihuana being associated with the down & out, the defiant...

Gravel-bars, riverbanks, scars
of the glaciers,
healing and nursing moraine—
tall hemp plants followed man

midden   dump heap   roadway slash

but the stories of Kerouac & Cassady on the road hadn't yet percolated out to Middle America...

To bind his loads and ease his mind
Moor to Spain, Spain in horse-manure
and straw, across the sea
& up from Mexico

—a tiny puff of white cloud far away.
we sit and wait, for days,
and pray for rain.

& though Gary Snyder's sage insights on tea & communion & hemp presaged the rucksack revolution of Dharma Bums, his poetry issued from no television sets.

The Atlantic Monthly

NOVEMBER 1966

Still three years from publication, Allen Ginsberg's article...

The Great Marijuana Hoax

First Manifesto to End the Bringdown

"I've never had a chance to explain my position on this subject without interruption, and to a large audience," says the poet Allen Ginsberg about the myths, the lore, and the legal aspects of marijuana. "So people mistakenly think I'm asking people to take dope-fiend dope."

became the first sane thing about the plant to be written this century according to the movement's Chief Archivist, the nation's first PhD in cannabis (& our favorite source) Dr. Michael Aldrich.

(Snyder misspelled his name in the dedication)

HOW much there is to be revealed about marijuana in this decade in America for the general public! The actual experience of the smoked herb has been clouded by a fog of dirty language perpetrated by a crowd of fakers who have not had the experience and yet insist on downgrading it. The paradoxical key to this bizarre impasse of awareness is precisely that the marijuana consciousness is one that, ever so gently, shifts the center of attention *from* habitual shallow, purely verbal guidelines and repetitive secondhand ideological interpretations of experience to *more direct, slower, absorbing, occasionally microscopically minute engagement with sensing phenomena.*

This early attack by Ginsberg launched the untangling of heroin from marihuana in the public mind & began the work of subverting the clumsy lies of Reefer Madness.

Ginsberg & Snyder saw mystifications in the cannabis smoke
just as the lab scientists saw in it their own mysteries.

On the marihuana question,
the report of Judge Prettyman's commission leaves us with a hint
in our hunt for the homunculus of old hemp.

Because it's still frightfully true what Kennedy
read in his interim report of 1963...

"Much with regard to the pharmacological phases of the problem is still being sought."

Or as Ludlow the Bard said
of the cannabinoid mysteries,

"Unfortunately, when we come to the chemical constituents of Cannabis, certainty is at an end."

The full identification of tetrahydrocannabinol (THC)
as the cannabis plant's psychoactive ingredient
resisted almost a century of chemists
trying to win the prize
& be the first to figure out
how this shrubby shrub
dominated the human history
of getting high.

The story of that chase
& its slippery spirals
almost starts to resemble
a search for the invisible
elephants in the room.

BUT BEFORE JUDGE PRETTYMAN'S FINAL REPORT COULD BE RELEASED...

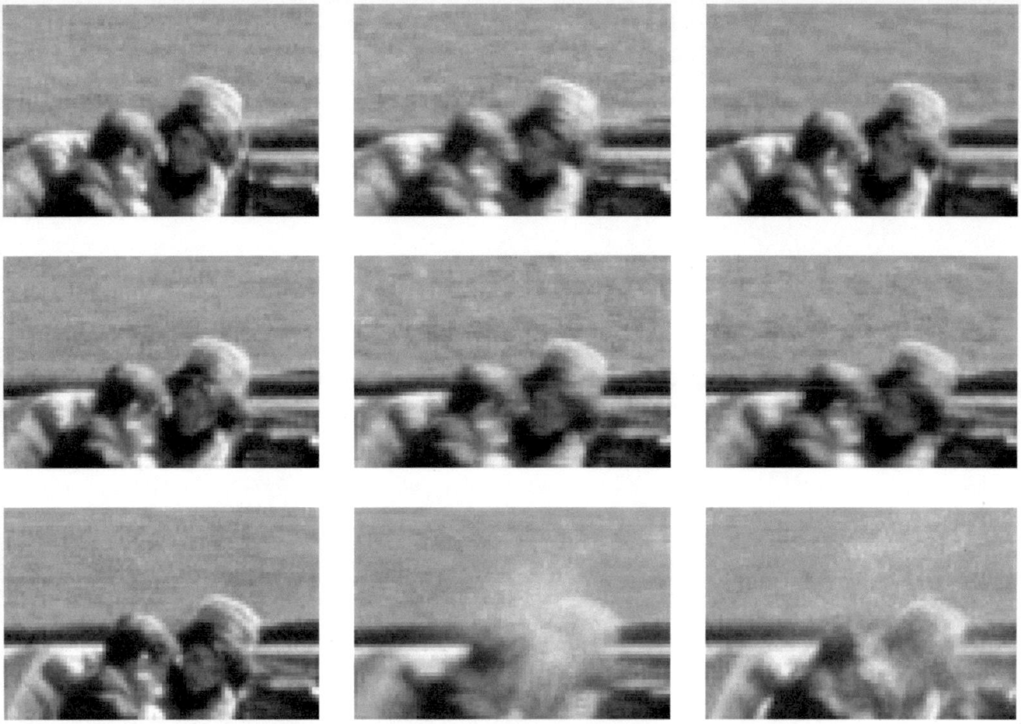

A SOMEHOW UNKNOWABLE NUMBER OF BULLETS

ENDED CAMELOT.

1964: THE YEAR OF THE STRUCTURE

As LBJ settled into his presidential power,
a bewildered nation stood aghast at the murderous politics sweeping Amerika.

Against these crimson curtains,
pot sparked back onto the scene
as a new stereotype slipped into
the ranks of teaheads:
college kids

With marihuana's grand advance into
the key demographics of uni students
& middle-class white kids,
the resulting media blitz served to
buzz cannabis back to center stage.

The Baby Boom caused by WWII made a fertile soil
for this new growth of hemp. In a surreal scenario,
half the US population was under the age of 30.

Marihuana found a new friend in this mass of adrift youth & forced a metamorphosis
in the anti-drug propaganda machinery maintained by the Federal Bureau of Narcotics.

Instead of the 'killer weed'
advocated by Harry Anslinger
in his mighty Reefer Madness
campaign of the '30s & '40s,
the rhetoric now shifted to
'dropout drug'...

well playing into Cold War fears of a nation of students & recruits
suffering from blunted ambitions & throttled motivations,

Mystery of the Strange Mexican Weed

American and Mexican Authorities Seek to Curb Growing Use of Dread Marihuana Drug That Stirs Its Victims to Atrocious Deeds of Violence

a new tactic in the long war

to demonize her always potent effects.

EXPOSING the Marijuana drug evil in swing bands

BY JACK HANLEY

Astounding facts about the habitual use of "reefers" among musicians

YOU'VE read about marijuana, of course. Newspapers recently have been full of stories about reefer-crazed young men and women jumping from windows, killing policemen, robbing, assaulting and running amuck, while under the influence of the

As Martin Lee puts it in 'Smoke Signals',
(the finest single work on the social history of grass)

"No single factor could account for why marijuana proved so attractive to large numbers of people on a continuing basis around this time. In some unexplained way, cannabis met the needs of young Americans as they grappled with 'growing up absurd' in a Catch-22 world...

Marijuana's status as a forbidden substance added to its allure. Smoking herb entailed an elaborate ritual – copping the weed, removing the seeds & stems (an antiquated folk art in the sinsemilla era), rolling a joint or packing a pipe & sharing it with others behind drawn shades while incense burned to mask the smell.

'When a young person took his first puff of psychoactive smoke, he also drew in the psychoactive culture as a whole, the entire matrix of law & association surrounding the drug, it's induction & transaction.

One inhaled a certain way of dressing, talking, acting, certain attitudes. One became a youth criminal against the State,' said Michael Rossman, a leader of the Berkeley Free Speech movement."

BUT,
AS THE STORY GOES,
WHEN THE SON OF A US SENATOR GOT BUSTED WITH GRASS, HIS PAPA WENT TO THE FEDERAL RESEARCHERS AT THE NATIONAL INSTITUTE OF HEALTH (NIH).
HE ASKED FOR THE FAKTS ABOUT THE HARMS TO THE BRAIN FROM CANNABIS.
AFTER A SHORT SEARCH & SCRAMBLE, THE CHIEF US BIOMEDICAL RESEARCH FACILITY WAS FORCED TO ADMIT THEY HAD NO RESEARCH ON FILE.

THE GREEN GRASS HIGHLIGHTED THE EMBARRASSING TRUTH:
DESPITE UNWRITTEN MILLENNIA OF HUMAN USE
& EXPERT FIDDLING BY MEDICINAL CHEMISTS
SINCE BEFORE THE US CIVIL WAR,
NO ONE EVER YET FIGURED OUT THE STRUCTURE
OF THE ACTIVE INGREDIENT STIRRING UP
ALL THOSE HEADTWISTING STATES OF POT.

In 1910 good old Fitz Hugh Ludlow followed in those worthy tracks with a quasi-mock rundown of the cannabinoid chemistry race in his 'An Essay On Hasheesh':

"I have extracted an alkaloid from hasheesh," says Preobraschensky, "& it is potent." "No, we have found the active constituent," says T. & H. Smith; "it is the resin cannabin." "No," says Personnne, "I have isolated the important ingredient; it is the amber-colored volatile oil, cannabene." "Oh, no," says Frankel, "I have discovered the active principle – it is a phenol aldehyde." "No, indeed," say Wood, Spivey & Easterfield, "it is we who have separated the only active ingredient – it is a red oil, cannabinol." "Oh, not at all," says Hamilton, "not one of these is the active constituent; in fact, the active constituent has not yet been isolated..."

Ludlow advises, *"In such an arena, where the masters dispute, it behooves the amateur to speak with a stammering tongue."*

As Razdan would roundly state 76 years later in NIDA's infamous Monograph #79, *"Both in man & animals effects produced by ingesting marijuana or delta-9-THC are unique. The gross effects produced cannot be classified pharmacologically as being primarily due to a stimulant, sedative, tranquilizer, or hallucinogen, although they share some properties with each of these."*

Sagacious researchers push back the veil of the unknown.

Members in good standing of the tribe of Science unite through curiosity & an abiding reverence for the Data. At their base, they are curious as cats – whom they often resemble.

But it's not for wealth or sex appeal that a chemist ends up at the laborious lab bench. They stay sitting there from a long-born lust to pull up just a little bit more about How It All Works.

What Paracelsus said of the alchemists is still true of their descendants, the chemists: "They are not given to idleness, nor go in a proud habit, or plush & velvet garments, often showing their rings on their fingers, or wearing swords with silver hilts by their sides, or fine & gay gloves on their hands; but diligently follow their labours, sweating whole days & nights by their furnaces. They do not spend their time abroad for recreation, but take delight in their laboratories. They put their fingers among coals, into clay & filth, not into gold rings. They are sooty & black, like smiths & miners, & do not pride themselves upon clean & beautiful faces."

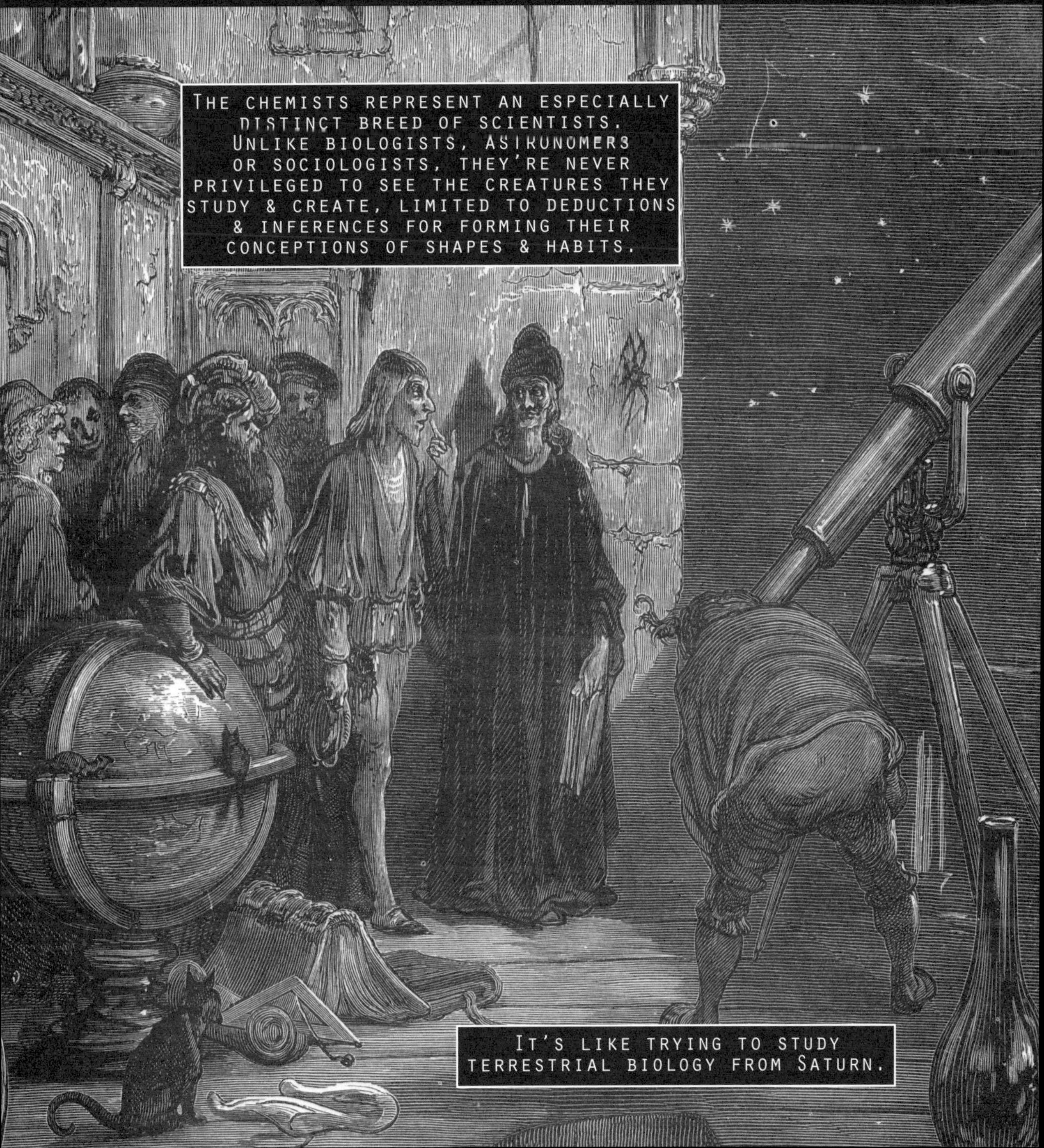

ONE NEEDS A MIXTURE OF ARTISTRY & DETECTIVE WORK
TO CRAFT THESE SCIENTIFIC HYPOTHESES
INTO IMAGINATIONS OF THE REAL.

TO CONCEIVE AN IMAGE IN THE MIND
OF WHAT CAN NEVER ACTUALLY BE SEEN,

THE CHEMISTS
CONNECT THEIR
REMOTE VIEWS
TO MAP OUT
THESE HIDDEN
REALITIES,

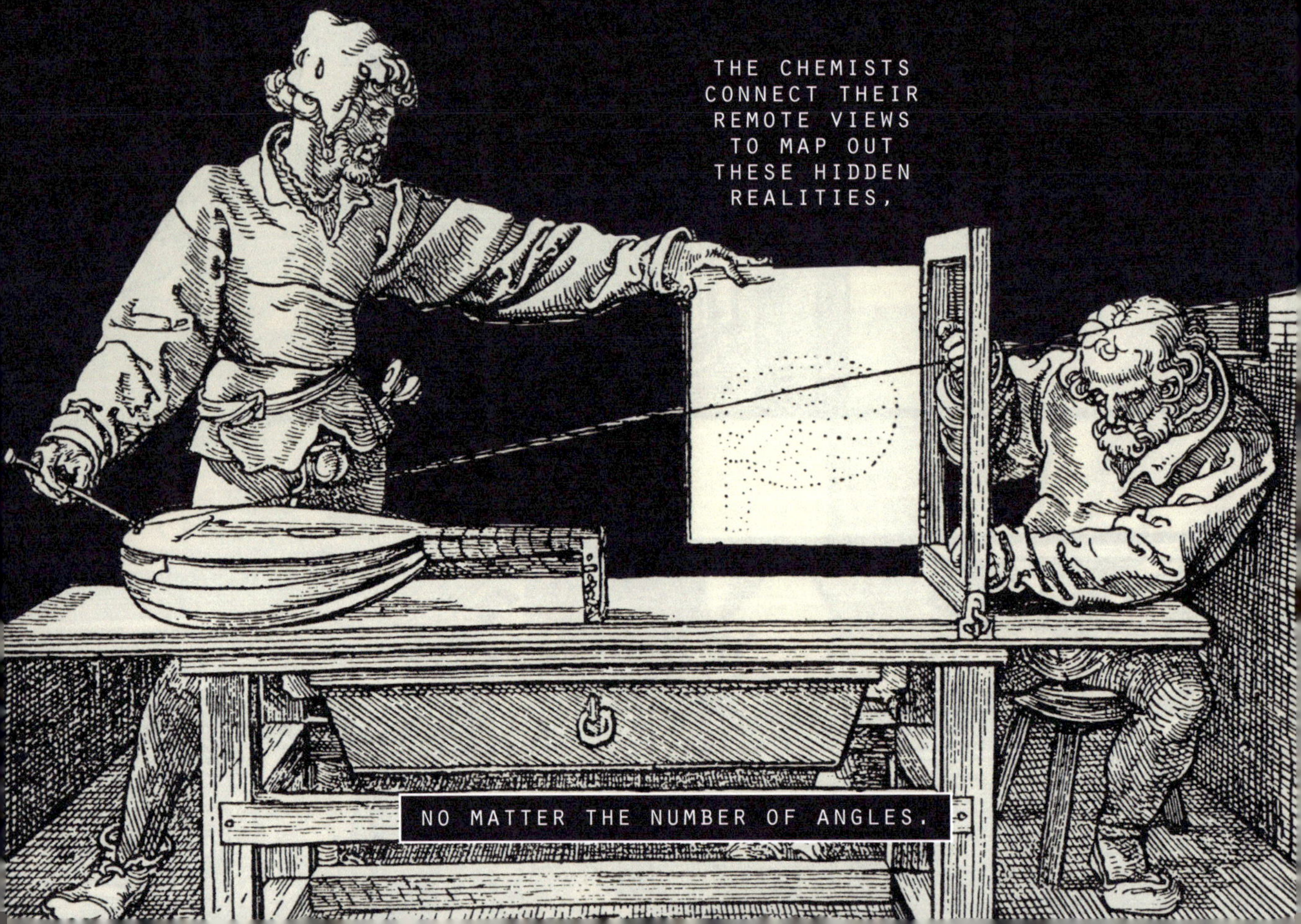

CHEMISTS CREATE THEIR INNER PICTURES
FROM pH STRIPS,

THEY VISUALIZE THE DANCING PATTERNS
OF ELECTRONS IN BOILING RATES,

THEY FORCE TELL-TALE DIVORCES &
JOININGS VIA THEIR ACID & BASE

& FROM THE CHROMATOGRAPHY
OF HEXANE EXTRACTS

THROUGH THE DARK VIOLETS
OF THE BEAM TEST,

THEY HOPE TO FILTER OUT
FROM THE GOLDEN SPAGYRIC

A CLONE OF NATURE'S CRAFTING

OR TO YIELD A NOVEL COMPOUND
NEVER BEFORE BEHELD ON THE EARTH,

THE DRIVE TO SEE JUST ONE MORE
CHEMICAL CREATION OF THE
10 MILLION VARIANTS
OUR SPECIES ALREADY
TITRATED & TAGGED.

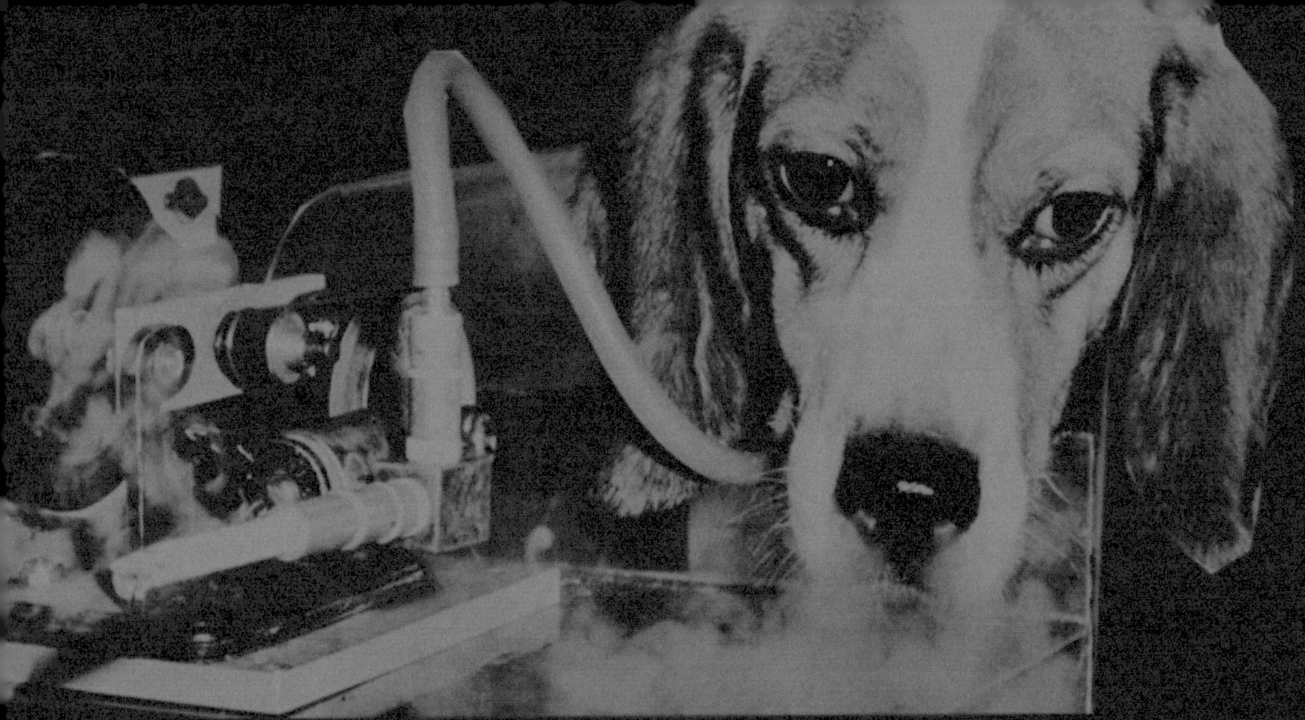

EVEN TODAY, AFTER OVER FIFTY YEARS OF CHEMICAL PRODDING,

WITH PAINKILLING PROPERTIES STUDIED IN TEN THOUSAND DOGS,

TOXICITY PROFILES
CRAFTED FROM A
100,000 MICE,

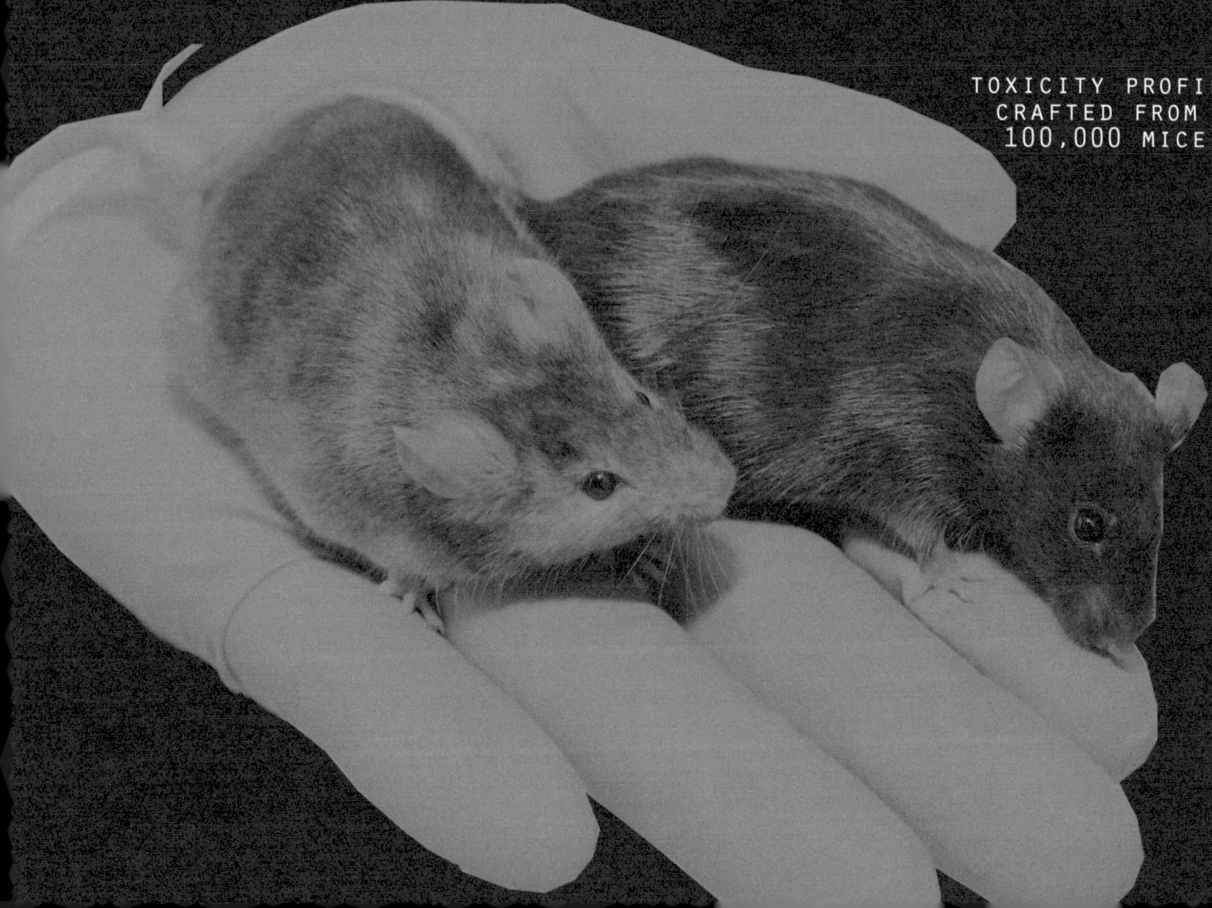

PHYSIOLOGICAL CHARTS
COLLECTED FROM THE DEEP POOLS
OF UNIVERSITY STUDENTS
ALWAYS HUNGRY FOR CASH,

& BRAIN SLICES FROM TROOPS
OF MONKEYS STRAPPED INTO
SMOKING MACHINES,

ALL CONTAINED
IN A TOWER OF
PEER-REVIEWED
LITERATURE SO
HIGH IT WOULD
TAKE YEARS TO
READ THROUGH
THOROUGHLY,

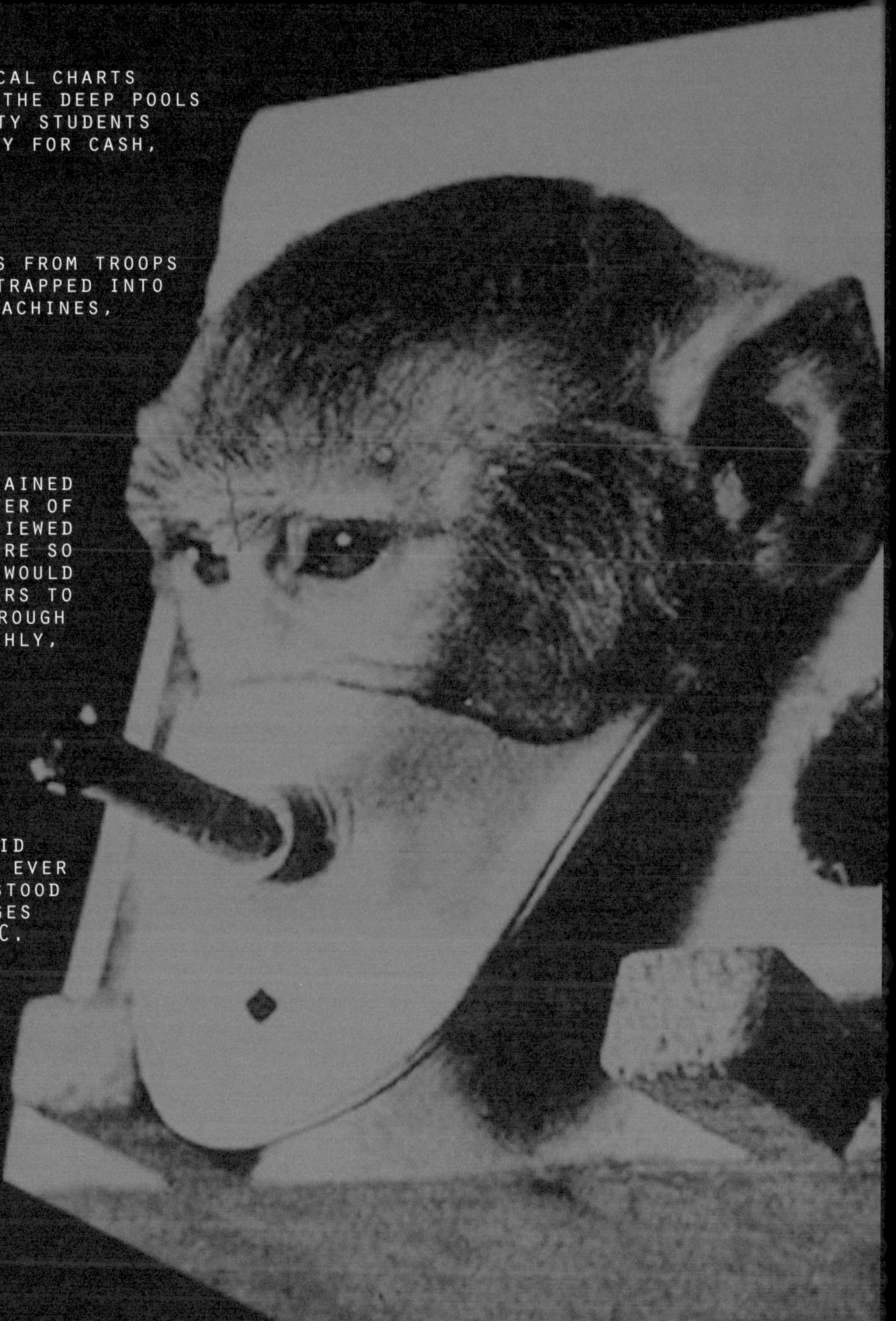

NO CANNABINOID
SCIENTIST WOULD EVER
SAY THEY UNDERSTOOD
ALL THE CHANGES
CAUSED BY THC.

BUT FOR THE STRUCTURE, THE DRAGON OF PROHIBITION HAD SPARKED
THE REVEALING FLAME.

NIH's embarassment at having no information about cannabis for
the Senator led to a stepped-up hunt for a researcher
to dig into the chemical components.

Since it remained difficult to study anything connected to cannabis in the labs of Amerika, an NIH official reached out to Dr. Raphael Mechoulam of Jerusalem's Hebrew University to see if he was still working with the plant. And though this would be his next step towards becoming known as the Father of Cannabis Research, he'd be the first to admire all the researchers who came before him.

And so many came before him.

Why did it take until 1964
– many decades after the structural solution of
morphine (the primary alkaloid in opium poppies) –
for the skeleton of THC to be laid bare?

Schultes himself pointed out,
the plants with psychoactive properties...

"may be divided conveniently into two broad groups:
those that contain nitrogen [N] in their structure
& those that do not.

Those with nitrogen are far more common. [The hallucinogenic compounds with nitrogen in their structure are alkaloids or related bases.]

The most important of those lacking nitrogen are the active principles of marihuana, terpenophenolic compounds classed as dibenzopyrans & called cannabinols – in particular, tetrahydrocannabinols."

Out of all the old plant medicines we threw
into our centrifuges,
mostly alkaloids emerged.

Their easy solubility in water & their readiness
for injection via syringe made them more
convenient to find & harness.

But THC's an
insoluble & glib
molecule who
resisted classification
& entrapment...

– a lipid –
– a fat –

a ghost that eluded
one hundred years
of chemical capture

& whispered
through the
keddahs

quietly
as an
elephant
on the
way to a
dance.

Because life for a chemist trying to study the cannabinoids felt like searching
for invisible elephants... on distant planets... using lousy telescopes.

How would a biologist track the mating habits of invisible elephants from space?
Especially when there're dozens of them...

BEARING STRONG FAMILY RESEMBLANCES, YET AN OVERWHELMING DIVERSITY...
THAT MAKES THEM A DEVIL'S TIME TO DIFFERENTIATE.

WHAT'S NECESSARY FOR THE CHEMIST WHO WANTS TO SEE?
MOST OF ALL, THE PATIENCE OF A FIELD BIOLOGIST'S STEALTHY SEARCH FOR THEIR SPECIES & THE MAKING OF CAREFUL DESCRIPTIONS FROM BEHIND THEIR BLINDS.
THE MINE IS ALWAYS BIGGER THAN THE GEM & AS THE FATHER OF EARLY CHEMISTRY JABIR IBN HAYYAN WROTE,
"The first essential in chemistry is that thou shouldest perform practical work & conduct experiments, for he who performs not practical work nor makes experiments will never attain to the least degree of mastery."
SO THE CHEMISTS SPEND LONG HOURS AT LAB BENCHES ON THEIR DILIGENT STALKS,

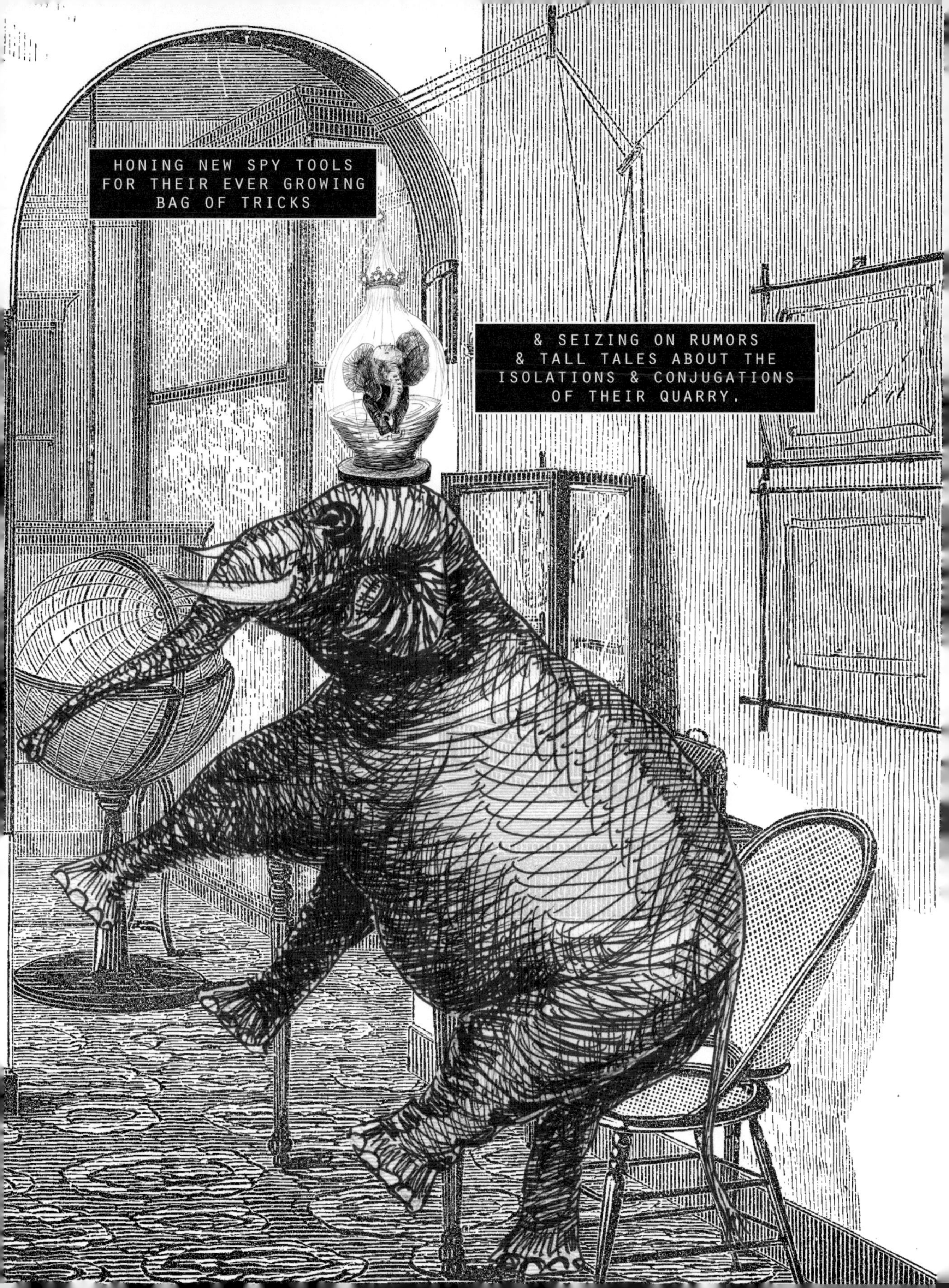
HONING NEW SPY TOOLS FOR THEIR EVER GROWING BAG OF TRICKS
& SEIZING ON RUMORS & TALL TALES ABOUT THE ISOLATIONS & CONJUGATIONS OF THEIR QUARRY.

The First Golden Age of Cannabis Medicine began after the US Civil War, ran strong until its first teetering spasms from the Pure Food & Drug Act of 1906 & then to its final death knell from Anslinger's Marihuana Tax Act in 1937.

but the uneven quality of Indian hemp shipped from the sunny Subcontinent hampered everyone's efforts. For the interested parties, one problem haunted their green studies: no reliable chemical reaction could be found for the presence of the molecules we now call cannabinoids.

As the revered Dr. Roger Adams laid it out:

"Between 1840 & 1895, most of the chemical investigations consisted in attempts to discover tests which would provide means of identifying a hemp extract. Numerous color reactions were reported, only one of which has received frequent application & until recently general acceptance. This is the so-called alkaline Beam test which is the purple color produced by treatment of a hemp extract with 5 per cent methanolic potassium hydroxide.

From our chemical results & from an extensive investigation of agronomic varieties of hemp by Dr. Matchett it can be concluded that this test is not indicative of a substance with marihuana activity."

i.e. they had only one test –
& no one knew it didn't work.

With variable starting materials,
a hodgepodge of recipes floating around,
& an assay wrongly viewed as trustworthy,

the early cannabinoid chemists surely lost
many hours screwing with the procedures
they found in the literature, cursing another
barren cycle again & again...

until you just can't help but wonder
if the author of the original paper
is a scoundrel...

or you yourself the dunce...

– with always the sneaking
suspicion that it's both.

So foremost to any scientist holding a jar of unknown:

How do you quantify what you've got?

Clinical and Physiological Notes on the Action of Cannabis Indica

How to know if you've got a chemical elephant in your cage?

Slip it into a bloodstream...

BY HOBART AMORY HARE, M.D.

& see if it swims, slops, stomps, rumbles or otherwise kicks up any kind of measurable ruckus.

Cannabis Indica has been before the profession for many years as a remedy to be used in combating almost all forms of pain, yet, owing to the variations found to exist as to its activity, it has not received the confidence which I think it now deserves. At present certain improvements made in the method of obtaining the extract from the crude drug have very materially increased its reliability, so that by selecting an article made by a responsible firm we may be fairly sure of

This report of 1887 – the earliest collected by Mikuriya in his landmark 'Marijuana Medical Papers' (besides the initial report of 1839 by O'Shaughnessy himself) showcases some of the potency assays for testing a distillation of Indian hemp on model organisms – animals from a wide range of sources & measured by a bewildering array of methods.

Dr. Hare began with large hypodermic doses of distilled cannabinoid oil injected directly into the leg of a frog.

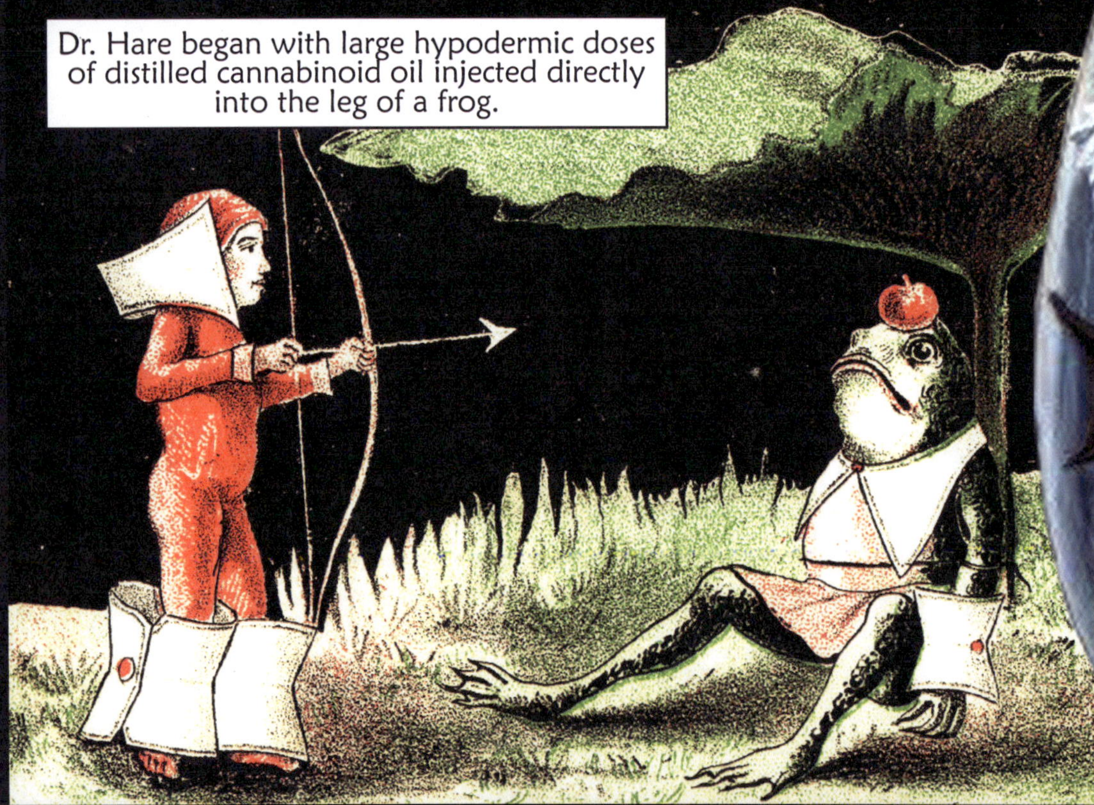

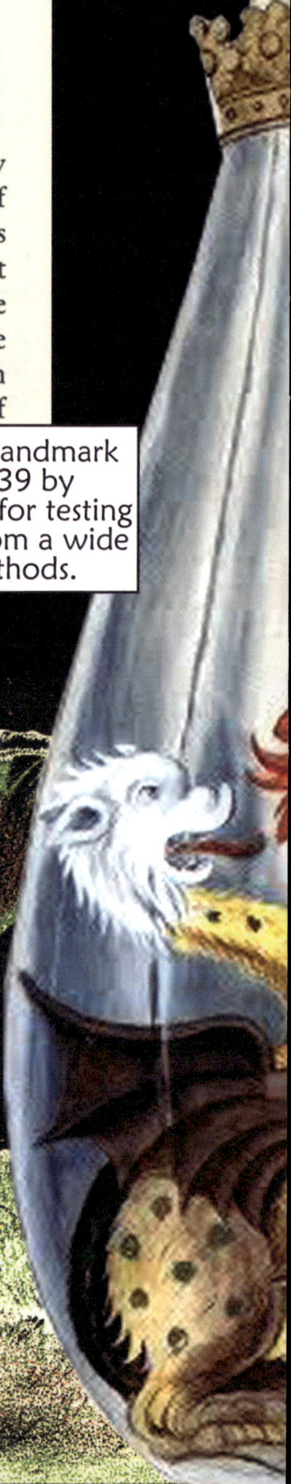

WEAR CELLULOID COLLARS AND CUFFS.
As expected, he found the application of intense heat to the untreated limb...
caused a powerful leaping reaction away from the noxious stimuli.
But the other leg, shot full of cannabinoids, could be totally burned off the body without eliciting a reaction,
the frog sitting oblivious to the removal of its leg by a red–hot poker.

Hare performed these frog tests in the hope of understanding why this herb helped his patients. Amidst the wider medical community, he reports that "within a few years this drug has become particularly prominent in its use in migraine."

From his own experience, he offers that it works particularly well on severe & intractable cases,

"not only in regard to the attack itself, but by acting as a prophylactic" to ward off future migraines.

He recommends constant small doses "in such a way that the patient is not conscious of any influence of the drug."

(That's the modern definition of microdosing & for all the rightful hype that gets paid to wee doses of LSD, ayahuasca or psilocybin, the microdosing of cannabis extracts is a vital health tool quite underutilized – especially for aging.)

Further, in the midst of a fierce migraine attack, Dr. Hare says you can't give too much. It must be "freely administered" until the pain & other symptoms greatly diminish since "so far no case of fatal poisoning from its ingestion has been recorded as occurring in the human being."

In his dog experiments, he found the drug so nontoxic that even 10 mL of Indian hemp extract injected directly into the jugular

produced nothing worse than a sound sleep

followed by bad dreams.

Indeed, within four hours the dog appeared to be waking up & showing signs of movement.

(Unfortunately, that experiment didn't end as planned because after the good doctor went to dinner, the lab assistant misunderstood his orders & killed the dog).

In a later experiment Hare finally found the lethal upper limit:

with 32 cc of cannabis extract, he could kill a dog of forty pounds.

That's over two tablespoons of the concentrated red oil right in the neck.

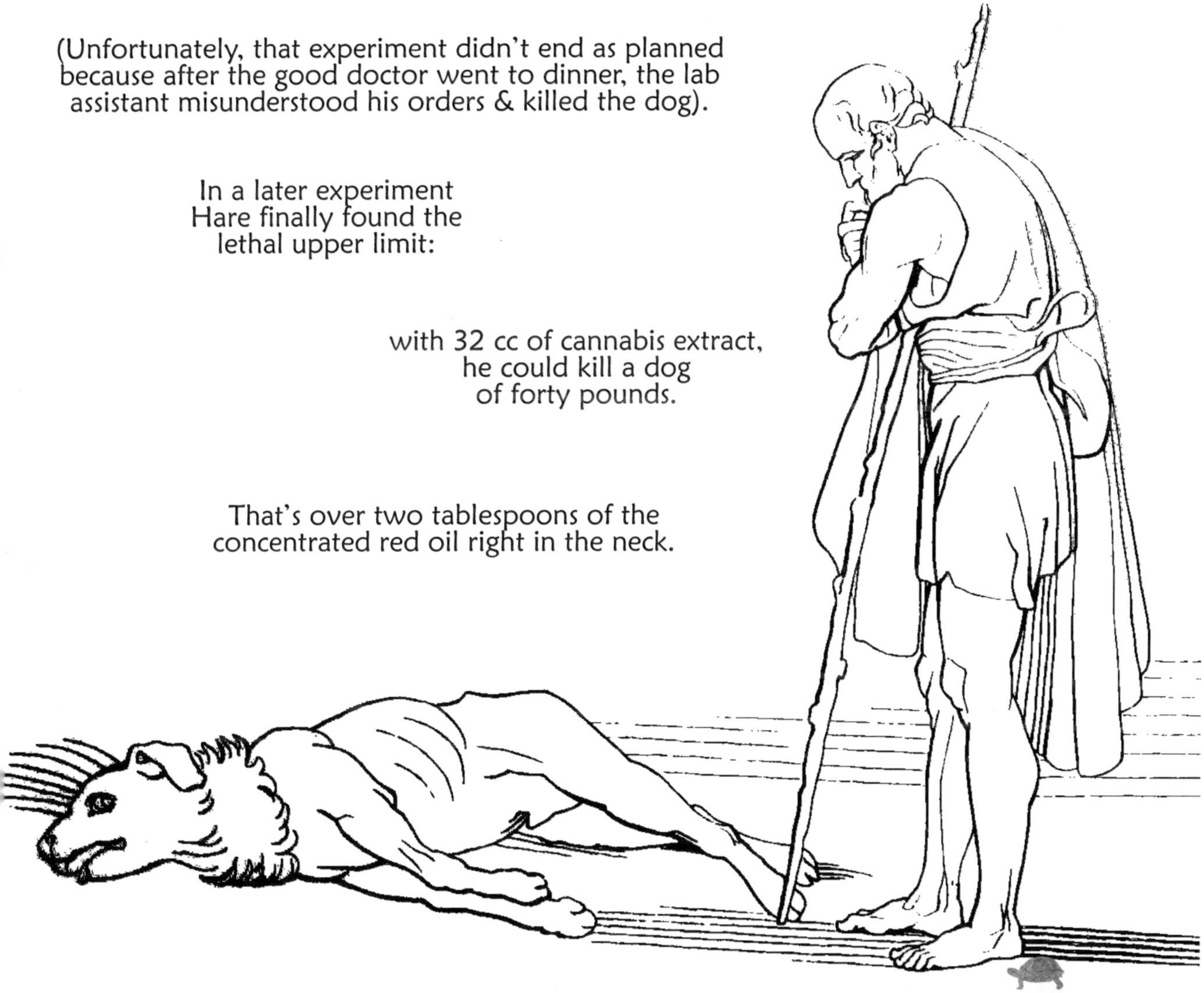

These scientists crave 'quantitative bioassays' – the potency of the drug cleanly correlated to a measurable physical effect.

Graphs. Basically, we want nice graphs.

In his 1898 report also included in 'Marijuana Medical Papers', Dr. CR Marshall assembled a thorough & quite sizable list of every time he got his dogs high.

He explained "physiologic investigations" & "experiments have been carried out in cats, dogs, rabbits & myself & investigations have been mainly confined to the administration of the various substances by the mouth – a condition necessitated by the comparative insolubility of the drugs."

In the paper he included this helpful flow chart on how to boil 2 kilograms of charas (cannabis concentrate) down into the potent 'red oil'.

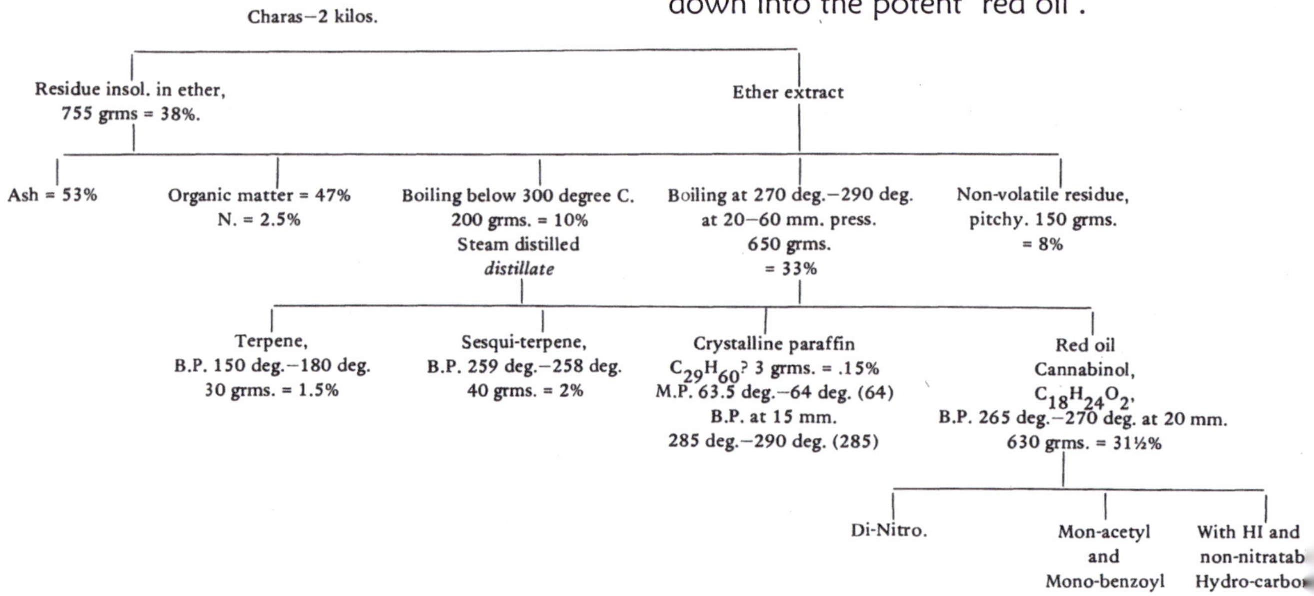

(Like virtually everyone in the century since, Marshall complained about the quality of the cannabis. In one sample, only 10% of the red oil turned out to be "cannabinol" – though as we'll see, it turns out he wasn't even looking for the right elephant).

Similar to how he separated out the red oil
by purification via sulphuric acid to disintegrate any alkaloids,
by trapping the cannabinoids under 22 mm of pressure
& by holding steady the boiling point between 265 degrees to 270 degrees C
in order to flush out the molecules with just the right characteristics,

Marshall also marshalled all the data he could think to measure
for determining the effects of dosing his dogs:
the pulse of their deep sleep, the slightly falling temperatures of their core,
their blunted responses to pain (how determined not indicated),
their reflexes (assumedly a doggy funny bone test)
& his suspicions of a depressed olfactory sense.

He made sure to note it all:

"Vomiting was a frequent symptom, salivation … a rare one.
In the Fox Terrier, increased micturition was occasionally obtained."

read: the nervous one peed
a little more.

Without reliable chemical tests, the field turned to bioassays – the measurement of effects in model organisms. As it evolved, the standard method to measure the amount of cannabinoids became the canine 'static ataxia' test – the dosage of the drug correlated to the loss of motor control in a dog.

According to Dr. Roger Adams,

"The determination of the cataleptic condition in dogs reported by Fraenkel, as accepted in former editions of the United States Pharmacopoeia for bioassay of Cannabis extracts, was developed further by Walton to attempt to make the effect the basis of a quantitative test.

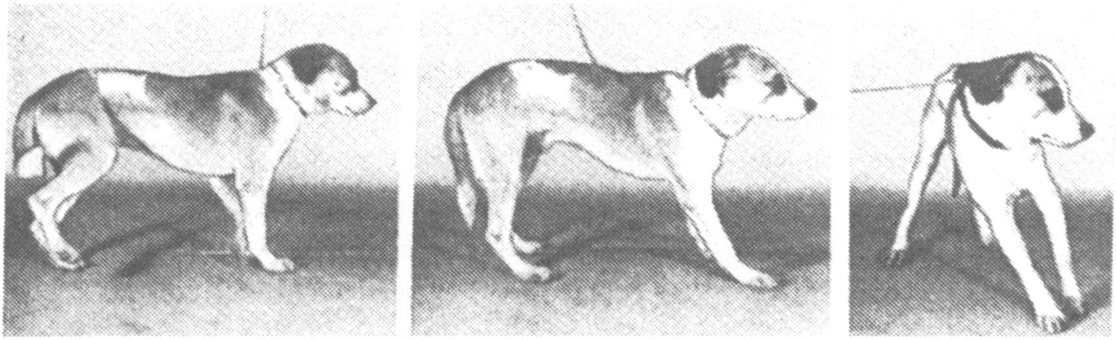

Six arbitrary stages of intensity of effects were recognized;
first a slight depression; second, a barely detectable ataxia [poor muscle control]; third, an obvious ataxia;
fourth, a marked ataxia in which the animal frequently pitches forward & barely catches itself;
fifth, inability to stand alone; sixth, inability to rise & plunge about.
The ataxia is chiefly a static one & is manifested particularly by swaying movements."

Besides other issues like large variability between groups & within individuals,

the high canine cost,

& the cost of high canines,

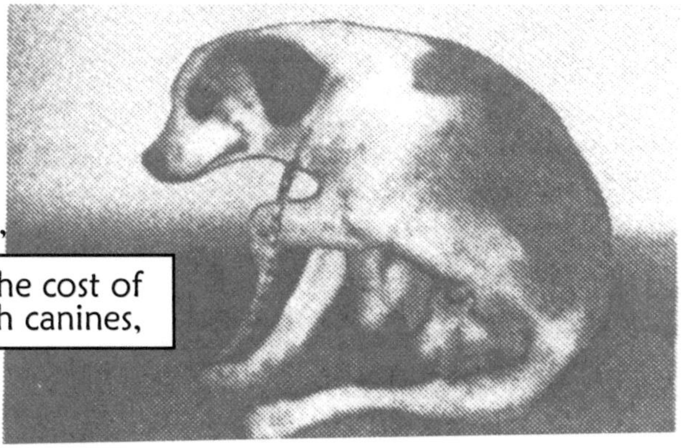

one problem loomed chiefly for the researchers always pressed for time: a single assay took at least a week to perform.

In an artificer's breakthrough, Dr. Walter Loewe brought his principle of 'Bioassay by Approximation' to the canine ataxia test. By performing tests in the same animal & applying different dosages, he raised the accuracy of the test up to a much more respectable ten percent. And now it took just one day to carry them out.

But these scientists were well aware of the truth.

There could be dozens of species of cannabinoids floating around in that red oil,

but if they didn't make a dog somehow twitch, tantrum, vomit or react, they'd leave no trace in the bioassays.

To measure the psychoactivity of a drug in humans, one can only use humans. It's for this attention to scientific duty that led Dr. Marshall to record 32 experiments of his own ingestion of various dosages & concoctions derived from the red oil.

The thought of this unfettered access might make modern stoners flush in envy but Marshall's entries seem to mostly chronicle unpleasant tiredness that occasionally resulted in a few days of hangover & mental exhaustion.

To experiment with drug interactions, he'd sometimes mix in opium or extracts of terpenes for good measure, a cocktail much harder to find this century than that

Dr. Walter Siegfried Loewe of Cornell must rank as one of the very few clinical chemists
to wind up on J. Edgar Hoover's 'enemies list' of disloyal Americans.

When Dr. Loewe first came to the attention of the domestic spy–in–chief,
his efforts to characterize the nuances of our still novel hormones
had already earned him wide respect as one of the world's top endocrinologists.

In Loewe's storied career,
he created the first urine test for testosterone–type hormones,
improved a similar test for the estrogen levels of ovulators
& was the first to correlate the peak of that hormone's cycle
to be the highest point of ovulation.

With over 50 published papers on this yet murky & unplumbed world,
Loewe became one of the first scientists to imagine the hormonal yin & yang
between our estrogens & our androgens,
a pattern he found repeated in the sex hormones of plants
(but since these make love over the course of a season,
it's only the occasional botanist, gardener or plant whisperer
who can remain still enough to spot their orgasm).

In addition, Loewe made early contributions to the radical hot new field of anesthesiology –
a terrain of medicine then without any precedents worn smooth into a set
of standardized procedures; but a process more like a delicate dance of dosages that required
patient observation, hard trials & alchemical diligence.

Until quite recently, when machines & continuous monitoring removed
almost all of the guesswork out of it, the practice of anesthesiology
was known as one of the trickiest of the medical arts.

So how did this Bavarian Jewish pharmacologist fleeing Europe
as the steel trap of the Final Solution sprang shut
manage to land on the disloyalty list of the
Amerikan Federal Bureau of Investigation?

Well, Loewe began annoying them early with gnattish swipes at the domestic drug warriors attending the 1938 conference organized by Anslinger's Federal Bureau of Narcotics.

Loewe argued bluntly with the commissioner that cannabis required much more basic research since even the active ingredients were still "chemically not known".

Much worse, after conducting a major survey of the issue of cannabis addiction in 1950, Dr. Loewe concluded:

"About habituation in the sense of addiction & tolerance, the cannabis agents appear to be among the 'drugs' in the last place."

He soon became one of the first researchers to look at the individual cannabinoids. He compared & contrasted the effects of cannabidiol (CBD), cannabinol (CBN) & the lab creation, Synhexyl (a molecule Sasha Shulgin labeled 'deservedly forgotten').

In Loewe's early experiments on monkeys, some of the first work done in non–human primates, he found the red oil to cause sedation, body sag & ptosis (a fatigued droop of the upper eyelids).

Loewe found that an injection of high levels of CBN into rabbits caused them to roll over within a minute & undergo violent convulsions until dead.

But there's never a guarantee that the results in model organisms

will turn out the same in humans.

No matter how similar all of us mammals might be, there are some ways that rabbits & rodents get high wildly different from us sapiens.

For instance, It appears that only in rabbits does the eyeball experience an anesthetic effect from topically applied cannabinoids.

Testing these weed eye drops for pain involves von Frey hairs – filaments of increasing size – pressed against the eyeball & then measuring the number of blinks.

For us, dear reader, smoked cannabis might lower the intraocular pressure for glaucoma or, as one ethnographer found, help Jamaican fishermen improve their night vision, but for eye surgery, it doesn't look like cannabinoid eye drops will be beating out pharmaceutical grade cocaine. Mama Coca wins this round.

So often in science, it's the toolmakers who unlock the giant bounds forward:
the microscopic lens of van Leeuwenhoek,
the double–boiler bath of that first alchemist Mary the Jewess,
& currently the CRISPR technique bursting into biochem labs across the planet, the hottest thing in genetics since the PCR inspired by LSD.
Because testing the constantly growing pool of cannabinoids & their synthetic cousins in dogs & monkeys quickly got expensive,
Dr. Loewe pioneered the bioassay that eventually grew into the field's standard for cannabinoid quantification: the mouse tetrad assay.
He turned to the laboratory mice just starting to be inbred & mass produced for scientific laboratories by William Castle & Clarence Little.
A standardization of life itself.

Loewe realized "the only sign of marihuana action common to all species of animals is a cataleptic effect". So that's what he honed in on. This condition of rigidly frozen muscles & a position fixed in place regardless of external stimuli or pain can even resemble death itself, inspiring Poe's terrible short story 'The Premature Burial'.

Loewe found that when using cannabinoids, a cataleptic state is "best manifested when the animal is placed prone upon an arrangement (brim of a beaker or parallel wires) for supporting it only at the thigh & jaws."

He found a direct & reproducible correlation between the dosage & the amount of freezing by the mouse in this precarious position (and also added the unique fakt that "excessive oral doses in mice produce diarrhea").

This catalepsy test was refined later into the 'ring test' by Dr. Roger Pertwee, author of the excellent review 'Cannabinoid Pharmacology: The First 66 Years'. Pertwee describes putting the mice across a horizontal ring held high above the floor. Distrustful of heights at the best of times, he found the mice tended to freeze in a dose-dependent manner.

In this freezing of fearful mice did the cannabinoid researchers finally find a reliable, repeatable test for measuring the levels of active ingredients in raw cannabis.

To finalize the Tetrad, Dr. Billy Martin's lab at Virginia Commonwealth University
added to the ring immobility test
3 more common bioassays for the measuring of:

ANTINOCICEPTION To test the painkilling power, one either employs...

the tail flick test or the hot plate test,

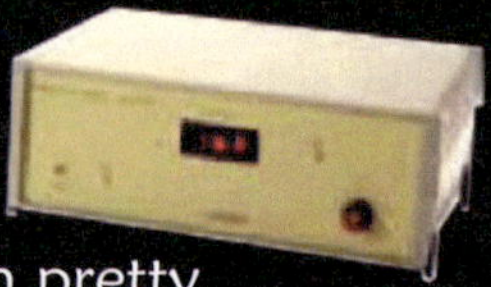

neither of them pretty.

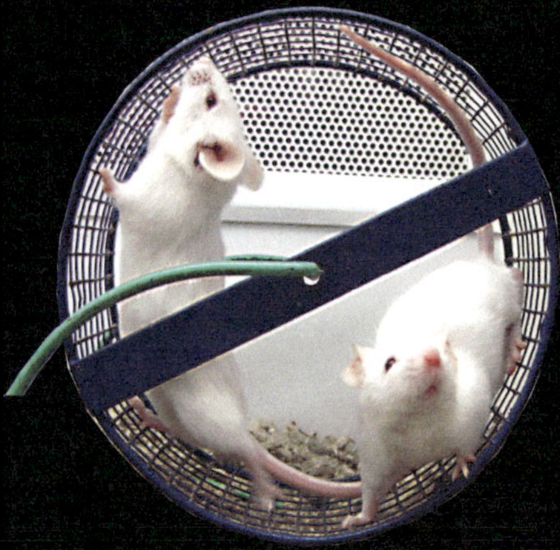

HYPOKINESIA

This test for the reduction of spontaneous activity gives a clue as to how grad students actually spend their time. For this one, they put potentially stoned mice in cages with floors marked by a grid of lines. Then the grad students count the number of times the mice cross those lines.

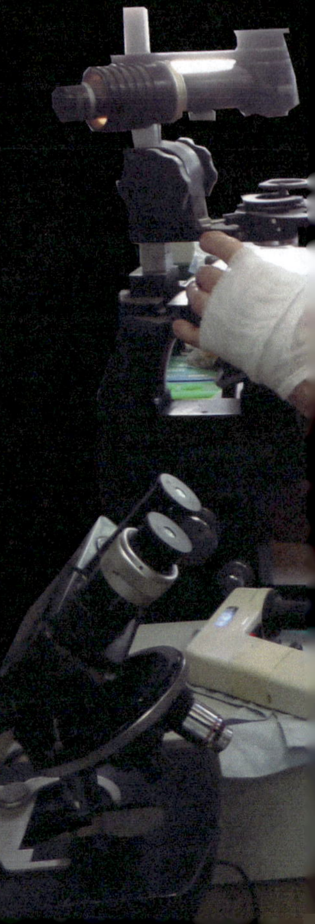

& HYPOTHERMIA

A rectal measurement of lower body temperature.

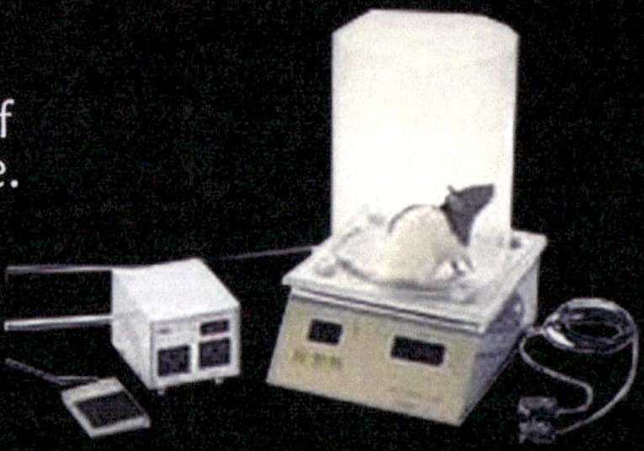

None of these tests individually predict for cannabis-like activity but if the model organism responds to all four, it likely – if not reliably – indicates the presence of cannabinoids.

Despite all the trauma imaginable from the previous pages, it must be definitively stated that many a lab scientist forms relationships with their lab animals – especially the intriguingly intelligent rats – & these researchers rue the harm necessitated by the Great Work.

(It's said that some PIs recommend to the young scientists they mentor to try, when possible, to conduct their research on human freshman & sophomores because one's less likely to get attached to them).

Dr. Ruth Galilly, one of the key scientists who worked with Dr. Mechoulam at Ein Kerem Hospital in the Jerusalem hills, well demonstrates this affection for the mice she utilizes in her experiments on the immune system.

Sometimes in the cold winters, she cheats & instead of keeping them in the barren animal care facility, she sets them up with a space heater in her own office.

Though now she's frustrated.

Decades of work have borne much fruit for her on CBD's power to lower inflammation & on its efficacy against an autoimmune disorder like diabetes – but why continue that line of experiments in animal models?

With her impressive mass of data, she's clearly proven that the non-psychoactive CBD is ready to be tested in humans. No more mice need be sacrificed to suggest that a basically nontoxic, readily available medicine should be forestalled any longer.

But she keeps on working, well past retirement, to keep pumping out the Papers that should be convincing the mainstream...

Later we will sit with Dame Galilly to hear more about her research. But for now, we just want to stress her example as a bench scientist doing her best to minimize the murine suffering from this work.

The ability to work in lab mice helped the field's growth & enabled chemists to more readily craft
just the kind of elephant they desired for the task at hand,

whatever that might be:
a painkiller that doesn't cross the blood brain barrier,
a radioactive ligand to track its travels through our body,
a probe for interactions in the gut
or some other cannabinoid marvel.
And so it was that CBN fell first "before Man's ever–questing search".

To study an elephant, you've got to catch one first...
because not every biologist has the luck of Aristotle to have been gifted, as some say, an Indian specimen for dissection by his far-ranging former pupil Alexander the Great.
For a chemist attempting to measure a molecule at those depths, it takes cunning & the right tools:
A sharp goad,
a clever mahout,

a keddah made of silica gel with an ether elution,
& a vicious determination to get that elephant into a cage.

So now, dear readers, we will attempt to really stretch the elephant's trunk...

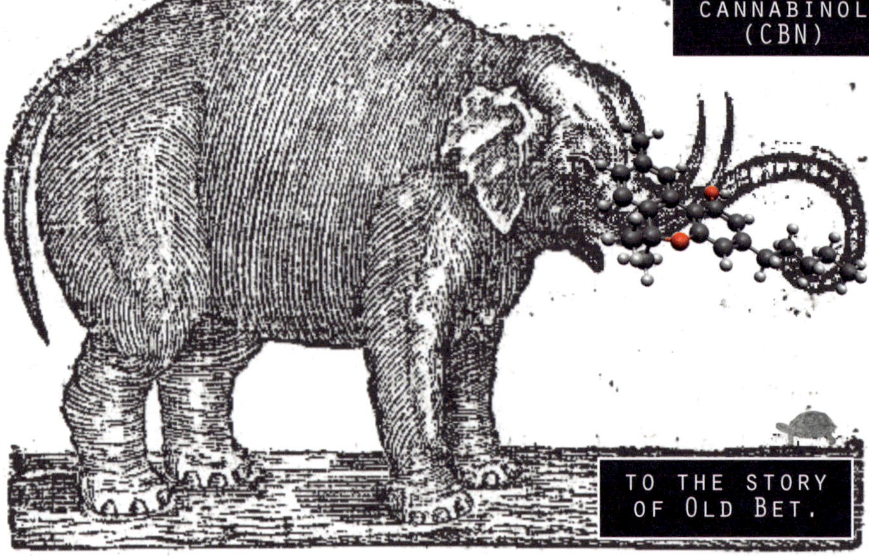

& align the search for cannabinol (CBN)

to the story of Old Bet.

THE

ELEPHANT,

ACCORDING to the account of the celebrated BUFFON, is the most respectable Animal in the world. In size he surpasses all other terrestrial creatures; and by his intelligence, makes as near an approach to man, as matter can approach spirit. A sufficient proof that there is not too much said of the knowledge of this animal is, that the Proprietor having been absent for ten weeks, the moment he arrived at the door of his apartment, and spoke to the keeper, the animal's knowledge was beyond any doubt confirmed by the cries he uttered forth, till his Friend came within reach of his trunk, with which he caressed him, to the astonishment of all those who saw him. This most c

Appropriately enough, both stories begin with shipments from India.

town, from Philadel-phia, where he will sta ... and weighs about 3000 weight, but will not hav ... 30 and 40 years old. He measures from the end of his trunk to the tip of his tail 15 feet 8 inches, round the body 10 feet 6 inches, round his head 7 feet 2 inches, round his leg above the knee 3 feet 3 inches, round his ankle 2 feet 2 inches. He eats 130 weight a day, and drinks all kinds of spirituous liquors; some days he has drank 30 bottles of porter, drawing the corks with his trunk. He is so tame that he travels loose, and has never attempted to hurt any one. He appeared on the stage, at the New Theatre in Philadelphia, to the great satisfaction of a respectable audience.

A respectable and convenient place is fitted up adjoining the Store of Mr. Bartlet, Market-Street, for the reception of those ladies and gentlemen who may be pleased to view the greatest natural curiosity ever presented to the curious, which is to be seen from sunrise till sundown, every day in the week.

☞ The Elephant having destroyed many papers of consequence, it is recommended to visitors not to come near him with such papers.

Admittance *ONE QUARTER OF A DOLLAR*——Children *ONE EIGHTH OF A DOLLAR.*

Newburyport, Sept. 19, 1797.

The first definitive date of an elephant ever seen on the shores of the United States is April 13, 1796...

But the adventure begins earlier on a Private Ship named, I kid you not,
AMERICA.

At St. Helena, off the coast of Africa, America reports taking an elephant on board.

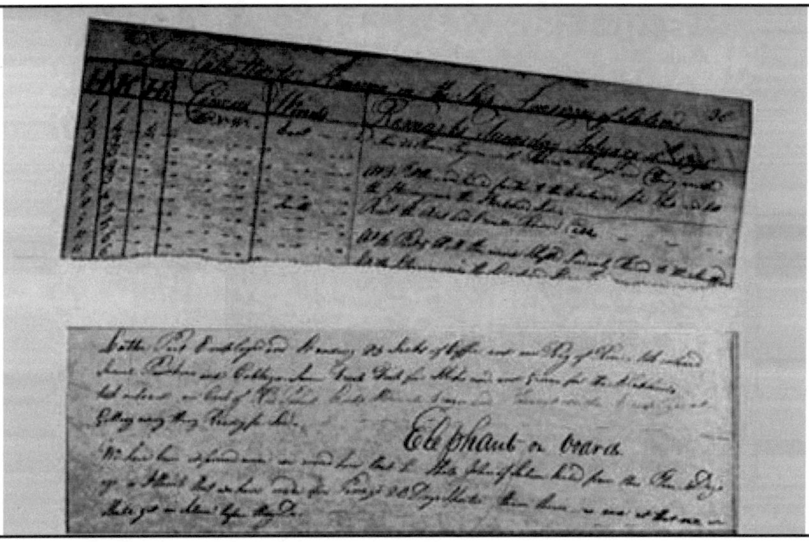
Elephant on board

The log follows the surely standard protocol of writing in much larger text when mentioning elephant boardings.

Even better, the writer of those words was none other than Nathaniel Hathorne, father of Nathaniel Hawthorne,

one of the first distinctly American authors to ever impress the wider world & whose dark romanticism caused John Greenleaf Whittier, our first great poet of pot, to esteem Hawthorne's "weird & subtle beauty".

Any reference to Nathaniel Hawthorne fires a whaler's blood because it's Hawthorne himself who served as Melville's inspiration for bringing forth Captain Ahab.

The crafty Herman suspected that no good Puritan could write such tales "shrouded in blackness, ten times black" without hiding some deep troubled soul secret of their own.

HAWTHORNE'S WRITING OFTEN CAPTURED THAT DRIVING OBSESSION OF THOSE MONOMANIACAL SEEKERS WHO PURSUE THEIR RESEARCHES AT ANY COST, AS SEEN BEST IN HIS DR. RAPPACCINI. A FIGURE OF LITERATURE WHO STANDS AS A LIVING MONUMENT TO THOSE STILL EXTANT SCIENTISTS WHO WILL STOP AT NOTHING FOR THE ACQUISITION OF ONE MORE FAKT:

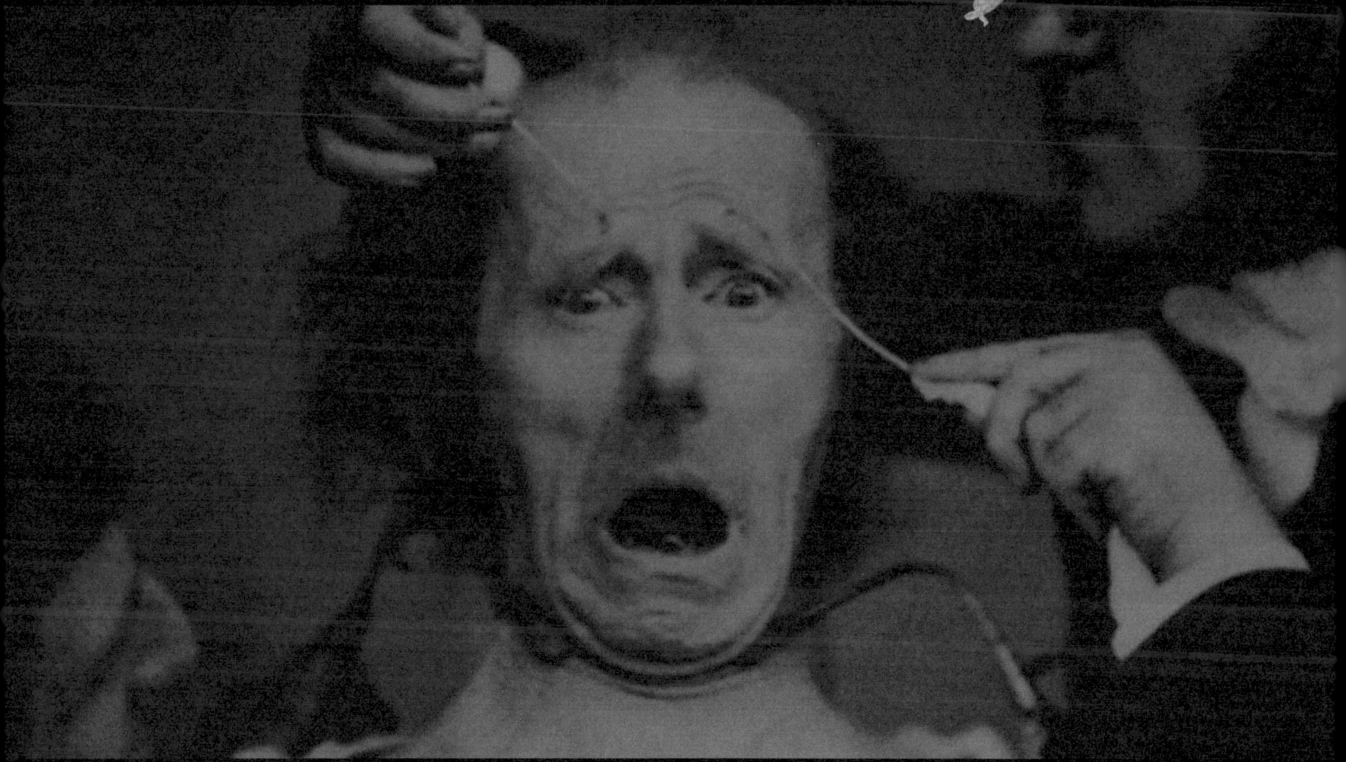

"But as for Rappaccini, it is said of him – & I, who know the man well, can answer for its truth– that he cares infinitely more for science than for mankind. His patients are interesting to him only as subjects for some new experiment. He would sacrifice human life, his own among the rest, or whatever else was dearest to him, for the sake of adding so much as a grain of mustard seed to the great heap of his accumulated knowledge."

HERE'S THE PINNACLE OF THAT RESEARCHER WHO WOULD THROW THEIR OLD MOTHER INTO A FIRE IF THEY COULD LEARN MORE ABOUT COMBUSTION FROM THE SCENE.

So what's the reason for transporting an elephant from Calcutta via those same Atlantic seas who hosted the unholy triangular trade of rum, sugar & enslaved people?

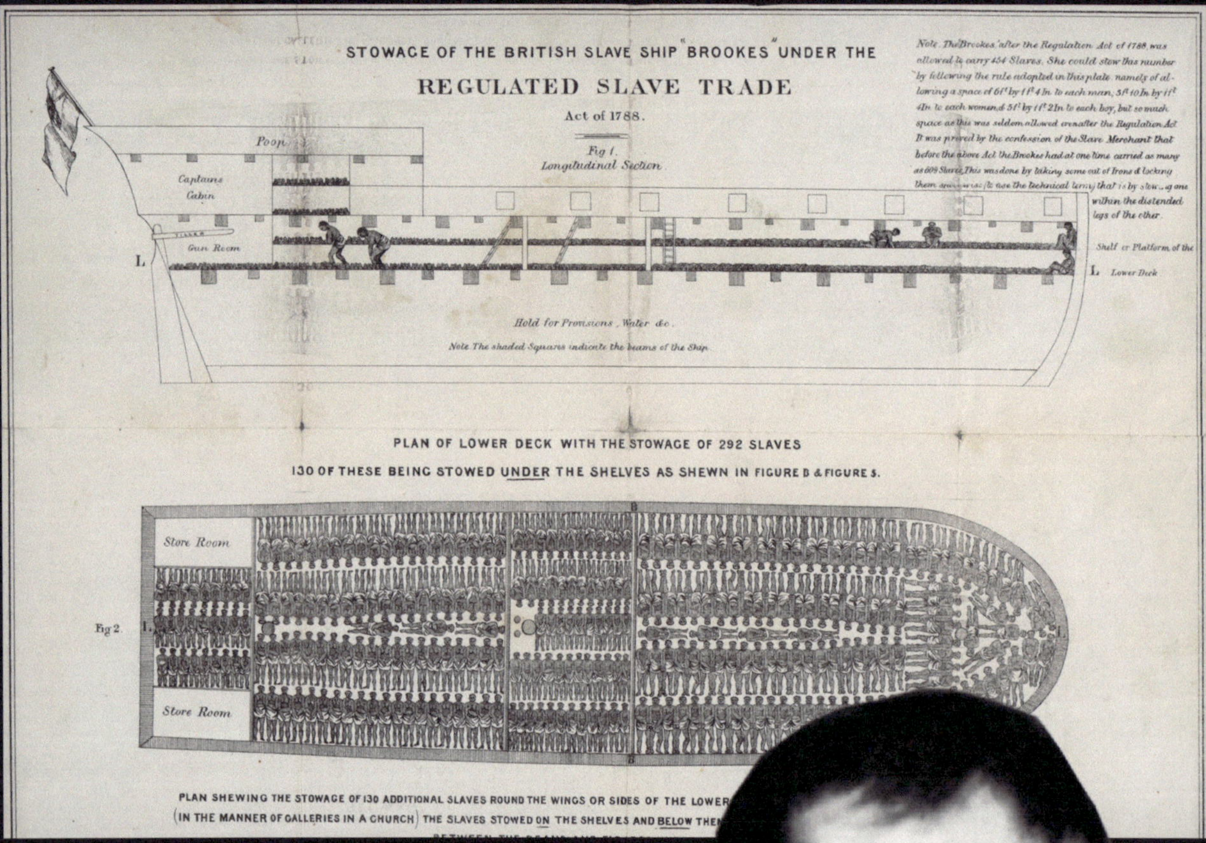

To make money of course,
the pursuit of that
Almighty Amerikan Dollar.

Captain Jacob Crowninshield of Salem, scion of a maritime trading family, wrote to his brothers from India:

"We take home a fine young elephant two years old, at $450.00. It is almost as large as a very large ox, & I dare say we shall get it home safe, if so it will bring at least $5000."

However, on the long crossing, she cut into his investment with her prodigious appetite.

When the water began running low, the Captain was forced to start serving her hogsheads of dark beer from the hold & from this, she developed a lifelong thirst.

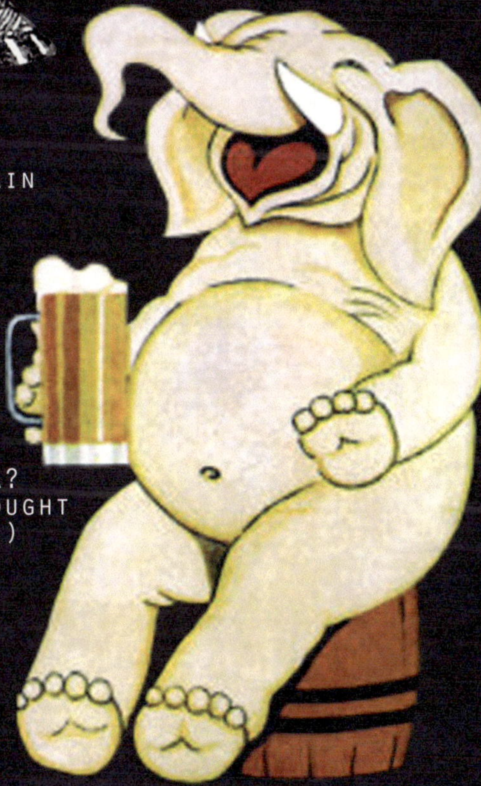

(A drunken elephant at sea? There's a sight I'd never thought my mind's eye would see...)

THE INITIAL WORK ON CANNABINOL
WASN'T QUITE AS MERCENARY
BUT ANY CLEVER CHEMIST
COULD IMAGINE THE FINANCIAL
POTENTIALS FROM SOLVING THE
SECRETS OF INDIAN HEMP,

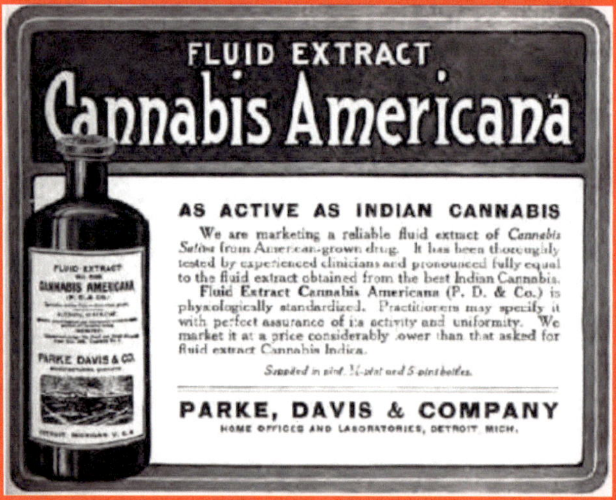

A MEDICAMENT DISPENSED BY THE
PHARMACEUTICAL HOUSES

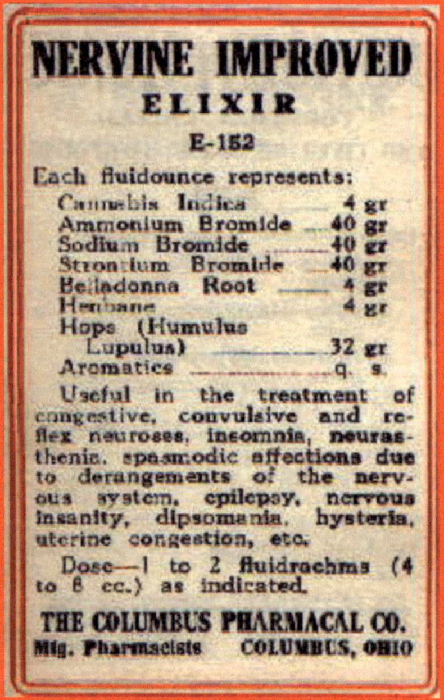

& A PLANT OFTEN ADDED
TO THE NOSTRUMS OF
PATENT MEDICINE
SALESMEN.

We cannot forget that a scientist possesses motivations as complicated & convoluted as any person & chemists are no less liable to be ignorant of potential profits than anyone else.

Hell, if you nab the right cannabinoid...

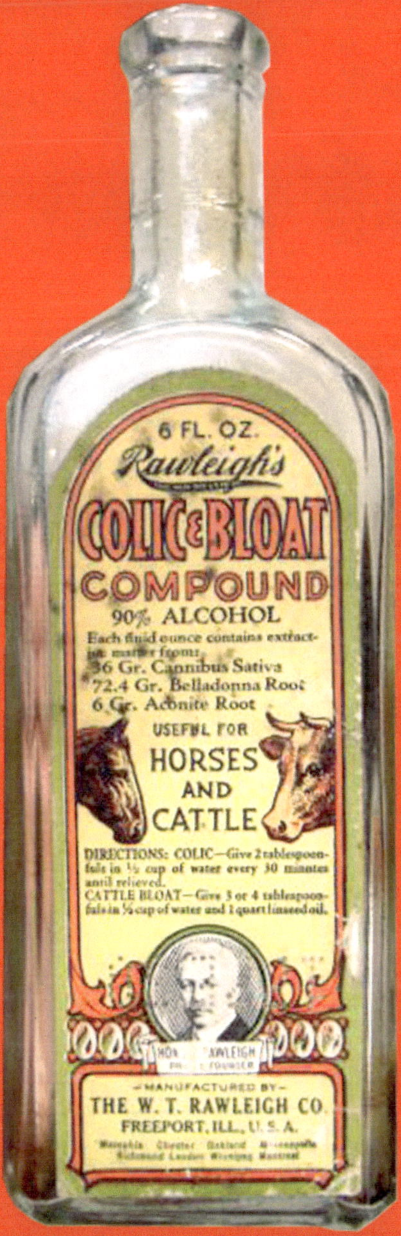

You might just corner the market for corns or colic & bloating in cattle.

For the elephant, the records show Captain Crowninshield, after making the initial bet of a 10 to 1 profit (as well as on the strength of an elephant to survive a sailing across the world), doubled that & made 20:1 after selling the two-year-old female for $10,000.

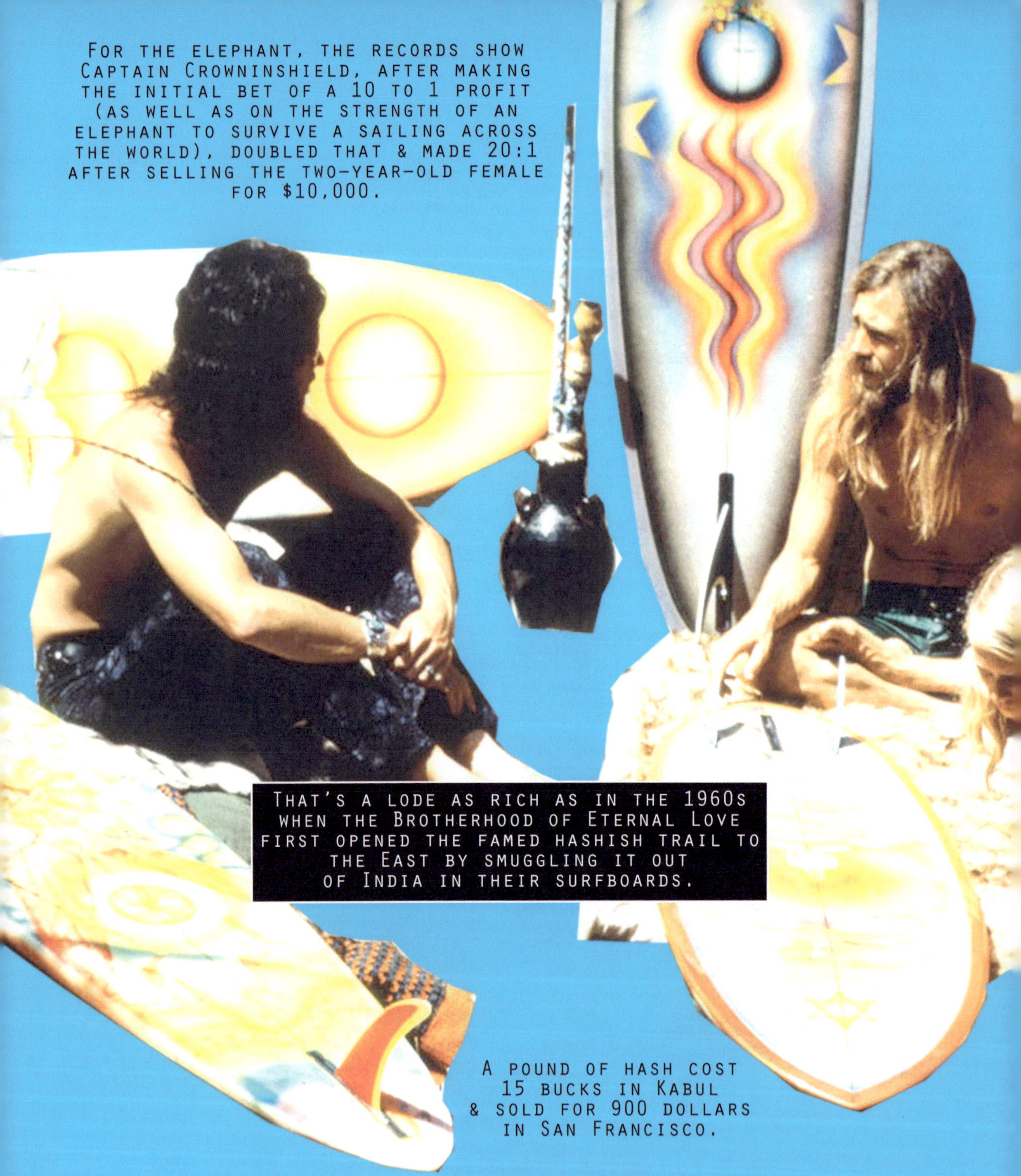

That's a lode as rich as in the 1960s when the Brotherhood of Eternal Love first opened the famed hashish trail to the East by smuggling it out of India in their surfboards.

A pound of hash cost 15 bucks in Kabul & sold for 900 dollars in San Francisco.

With messianic zeal, the Brotherhood poured those monumental profits back into cheaply spreading their than novel sacrament: LSD.

When the first elephant landed in the Land of Opportunity, her owners set her up on the New York City corner of Broadway & Beaver.

With an admission of two bits a gander, all could come to see her take a loaf of bread from your hand or mischievously steal important papers from people.

She spent over a dozen years on the road making silver for her owners...

because who of ye wouldn't travel far & pay dearly to witness an elephant uncork & down a dozen Guinness?

The road to CBN's characterization also took its tithe of years to complete. Much blood, sweat & sacrifice went into the hunt for cannabinol's structure. For the summary of CBN's early history & the many steps from her first spying to the final cloning, we turn to Dr. Roger Adams...

because if Dr. Mechoulam reigns as cannabinoid chemistry's guiding light

for the second half of the 20th century,

Dr. Adams performs the same function for the first.

And they both strike one as swell guys.

Dr. Adams summarized the early history of the cannabinoid field in this excellent lecture after he spent a few years on the problem at the request of the Federal Bureau of Narcotics.

BULLETIN OF
THE NEW YORK ACADEMY
OF MEDICINE

NOVEMBER, 1942

MARIHUANA*

ROGER ADAMS

Professor of Chemistry, University of Illinois

Harvey Lecture, February 19, 1942

THE facts on marihuana which I shall present to you this evening comprise the results of the coöperative efforts of three laboratories—the chemical investigations at the University of Illinois, the pharmacology at Cornell Medical College under the direction of Dr. S. Loewe, and the clinical experiments at Welfare Island Hospital under the auspices of the Mayor's Committee on Marihuana and under the immediate direction of Dr. Samuel Allentuck. All three laboratories acquired their supplies of raw materials from Dr. H. J. Wollner of the Narcotics Laboratory of the Treasury Department, and received much encouragement and stimulation from him. Dr. J. R. Matchett of the same laboratory contributed significantly to the general problem by devising a method whereby a fraction of the red oil of hemp, containing a very high concentration of the active principle, could be obtained.

Cannabis sativa, more commonly called hemp, is of peculiar interest.

He shares,

"In 1895 Wood, Spivey, & Easterfield were able to isolate from "purified red oil" [Indian hemp extract] by means of a treatment with acetic anhydride, a crystalline acetate, which was removed from the residual oil & which could be purified in the normal way. Hydrolysis of this pure acetate resulted in a homogeneous viscous oil which these investigators called cannabinol [CBN]. Until about a decade ago, when the fact was shown to be erroneous, cannabinol was accepted as the active principle of hemp."

That last simple sentence,
stripped of its scientific despair,
implied decades of frustrated chemists
trying to replicate results from the few others in the field
& months of wasted effort tinkering with processes
trying to trap an elephant that wasn't there.

NOT TO SAY CBN ISN'T SLIGHTLY PSYCHOACTIVE
BUT IT SURELY DOESN'T GET YOU TWISTED LIKE THC.

WOOD, SPIVEY, & EASTERFIELD
THOUGHT THEY HAD THE ACTIVE
INGREDIENT CORNERED...

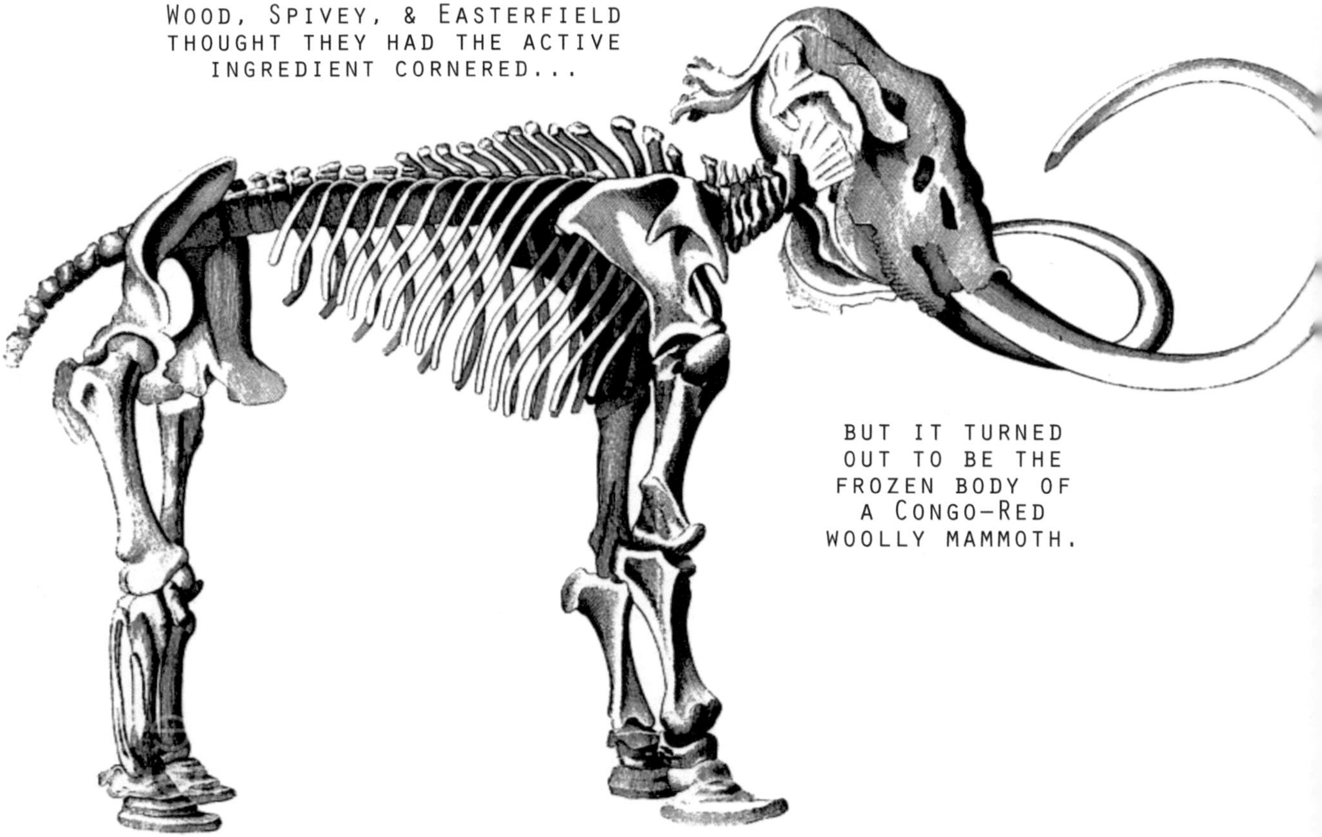

BUT IT TURNED
OUT TO BE THE
FROZEN BODY OF
A CONGO-RED
WOOLLY MAMMOTH.

THE TRIO HAD INDEED BEEN THE FIRST TO ISOLATE
A PURE CANNABINOID FROM THE PLANT BUT THEY NEGLECTED
TO TEST IT IN ANIMALS FOR SIGNS OF PSYCHOACTIVITY.

DESPITE NOT BEING ABLE TO MAKE OUT THE STRUCTURE OF CBN, AT LEAST THE THREE CHEMISTS COULD GIVE THE NEW MOLECULE THE NAME OF 'CANNABINOL' BEFORE THEIR WORK CEASED...

IN A MANNER THAT TOO WELL FORETOLD THE DARK DUSK OF OLD BET.

According to Walton's 'Marihuana – America's New Drug Problem',

"Wood, Spivey & Easterfield, the Cambridge chemists, were not able to complete their program because of a series of tragic accidents.

Wood barely escaped with his life when he took some cannabinol at the time he was preparing zinc ethyl. He lost consciousness, the zinc ethyl ignited & he was rescued from the burning room only with much difficulty.

Easterfield was killed by a violent explosion while attempting to hydrogenate cannabinol.

Spivey similarly perished while engaged in a synthetic study of the nitro-cannabinolactone."

May Aristotle bless all those scientists
who go down with their ships
& may we always remember those lost
while holding up a fluttering candle
against the dark heart of that Almighty Unknown,
enduring their diligent voyages
while seeking to navigate
by the Scientist's Prayer:

"God give me unclouded eyes
& freedom from haste.

God give me a quiet & relentless anger
against all pretense & all pretentious work
& all work left slack & unfinished.

God give me a restlessness
whereby I may neither sleep
nor accept praise
till my observed results
equal my calculated results
or in pious glee
I discover & assault my error.

God give me the strength
not to trust to God!"

– Steinbeck's Dr. Arrowsmith

Despite these tragedies & setbacks, the cannabinoid work continued.

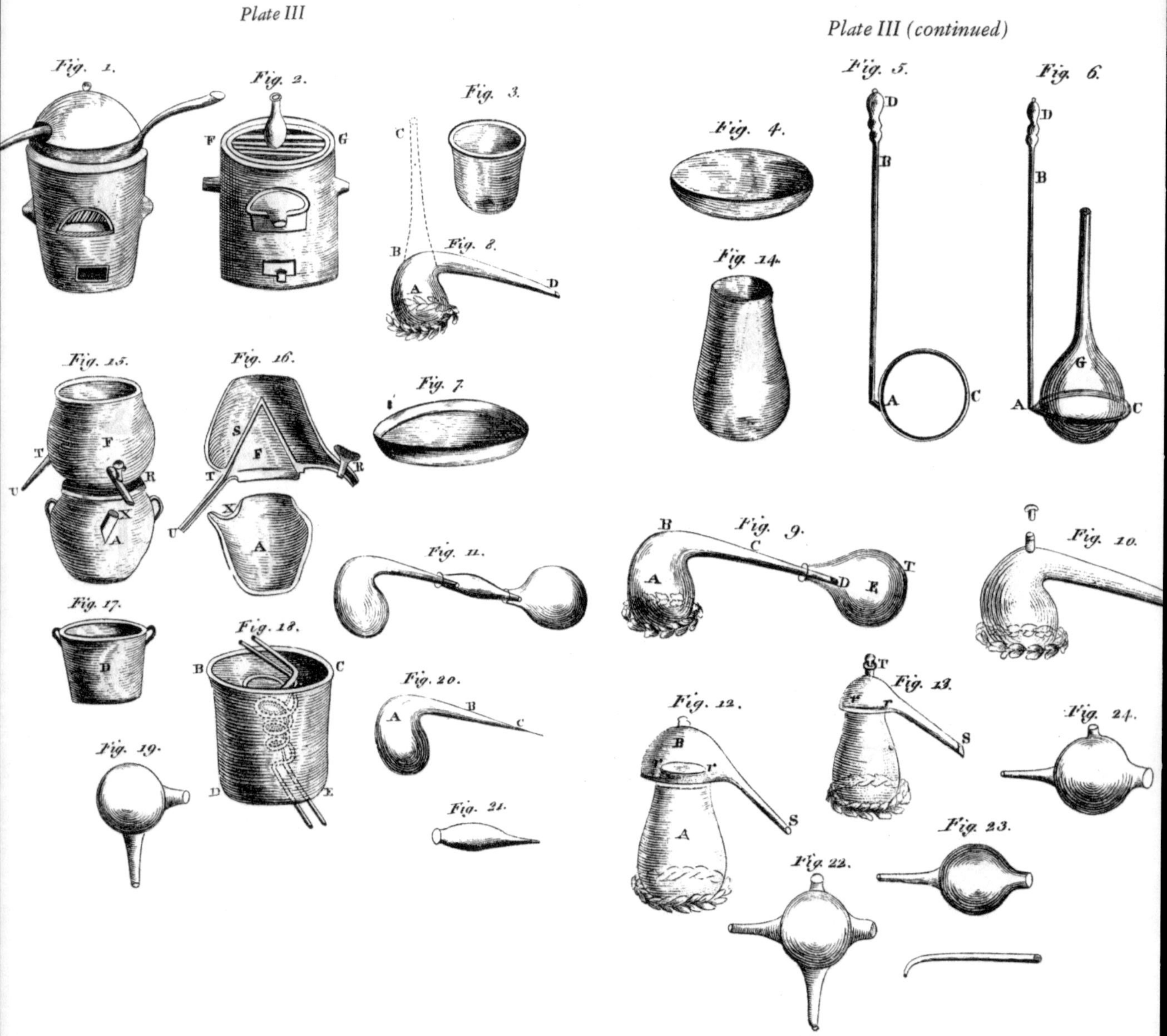

From there, the first step on the stairway spiraling up towards CBN's characterization was the elemental analysis,

– an experimental tool first developed by Antoine Lavoisier, the father of modern chemistry –

to find the fractions of carbon, hydrogen & the other heteroatoms like oxygen, nitrogen, sulfur & iodine.

Elemental analysis
usually happens
via combustion.

Burn your
sample to set
it free.

Throw Rappacinni's mother over the Bunsen burner
to see what comes off.

For your elemental analysis, heat the unknown cannabinoid compound in a bath of Mary.
Use all your artifice to lay a series of traps for collecting ALL the carbon dioxide, water & nitric oxide that boils off.
Weigh all the fractions to find the quantities of these familiar & known elements.

Keep scribbling out molecular formulas until you find one that yields a ratio of atoms adding up to your starting mass within the 0.4% of accepted deviation.
You're helped in this mathochemical monkeying because atoms only bind one to one, yielding proportions of whole numbers
A discovery in itself that took a few centuries of fiddling to theorize, hash out & finally codify into a Natural Law from the early precipitations of the Arab alchemists onward.

This business of elemental analysis is such a tricky & fascinating process that when Dr. Fritz Pregl's invention of methods for the micro–analysis of organic compounds took the 1923 Nobel Prize for chemistry,
at the award ceremony in Stockholm, the chemist delivered a full lecture on lab techniques to the King of Sweden & his flock of potentially perplexed diplomats & dignitaries.
Thus, with restraint, we will report simply that Dr. Marshall found CBN to consist of
18 CARBONS,
24 HYDROGENS
& A LOVELY PAIR OF OXYGENS...

ALL STRIVING TO SHARE THEIR ELECTRONS
& FIND THAT BALANCED HARMONY ALL ATOMS SEEK.

However, we're forced now to report that later work determined CBN's molecular formula to actually be: $C_{21}H_{26}O_2$

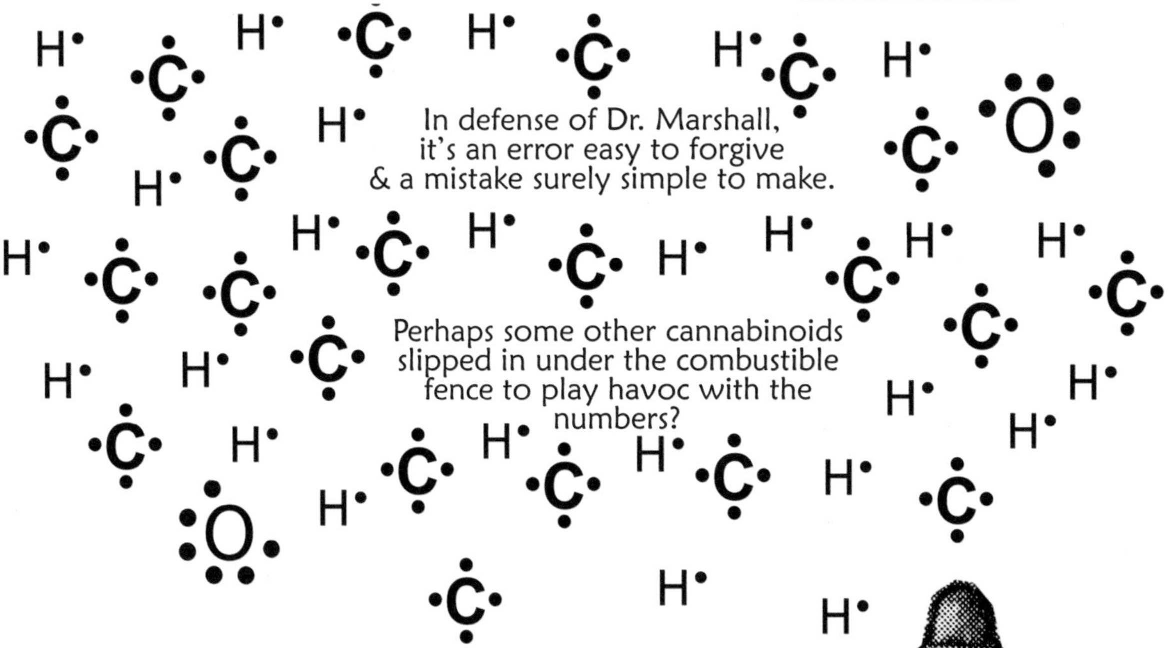

In defense of Dr. Marshall, it's an error easy to forgive & a mistake surely simple to make.

Perhaps some other cannabinoids slipped in under the combustible fence to play havoc with the numbers?

With THC weighing almost the same – just 4 more hydrogens than CBN –

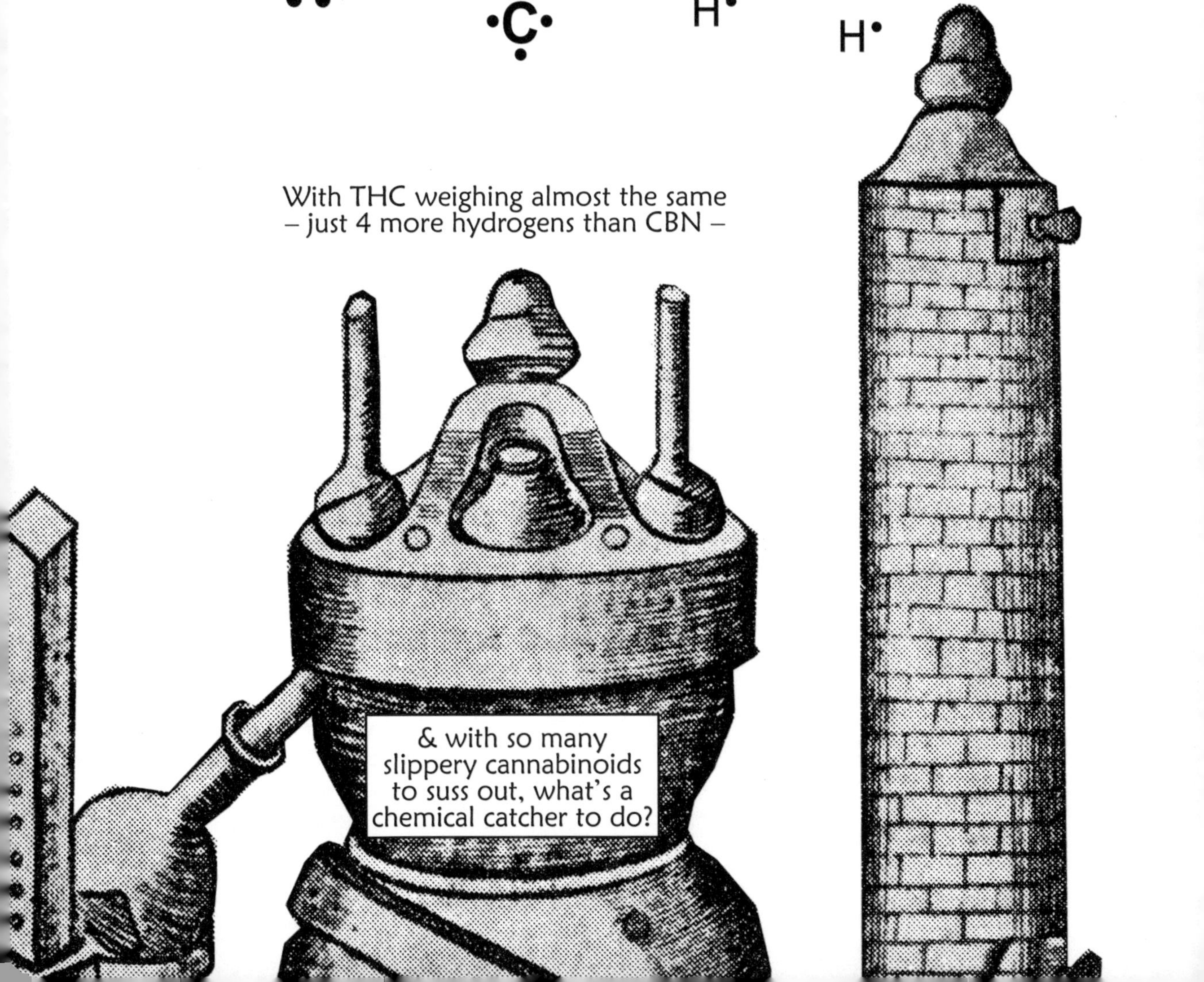

Similar problems plague the pachydermic historians
trying to make heads or tails of the situation.

According to the scant paper trail assembled to trace the First Elephant through Amerika, there's no proof that the female who sailed in past Coney Island was the one & same Old Bet...

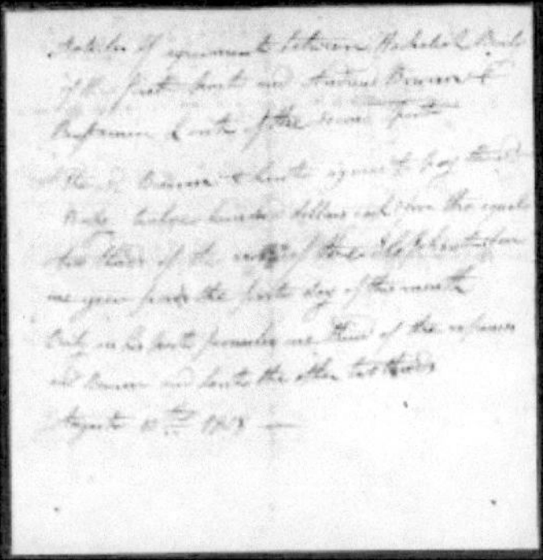

whose name first appeared in this bill of sale on Aug. 13 1808 by Hachaliah Bailey, the founding pioneer of the modern circus.

Whether this was the elephant that crossed on the good ship America led to enormous confusion...

December 17, 1808.

Property Somers Historical Society

A LIVING ELEPHANT

TO be seen at the house of Wm. Satterwhite from Tuesday the 20th to Saturday the 31st instant. Those that wish to gratify their curiosity by seeing the wonderful works of nature, will do well to call previous to that time. Perhaps the present generation may never have the opportunity of seeing an Elephant again, as this is the only one in the United Sates, and perhaps the last visit to this place.

The Elephant is eight years old and is upwards of seven feet high.

☞ Admittance 25 Cents. To be seen from 9 o'clock in the morning till sunset.

44 Dec. 12.

& many letters of correction posted to scholarly journals & popular historical magazines by earnest elephanteers who wrote in to correct those who erred about these early fakts.

Before Old Bet, Bailey's traveling show contained only
4 wagons, a horse, a few pigs & a trained dog.
She fulfilled that circus pioneer's dream:

For the first time in Amerikan history, a tradition began where you could go see
a real live Behemoth under the Big Top.

Just as the physical presence of elephants in the States opened them up to further study & elucidation,

in a like manner did Dr. R.S. Cahn finally (mostly) figure out the structure of CBN in the early 1930s

& become the first chemist to get an (almost) complete picture of the structure of a cannabinoid.

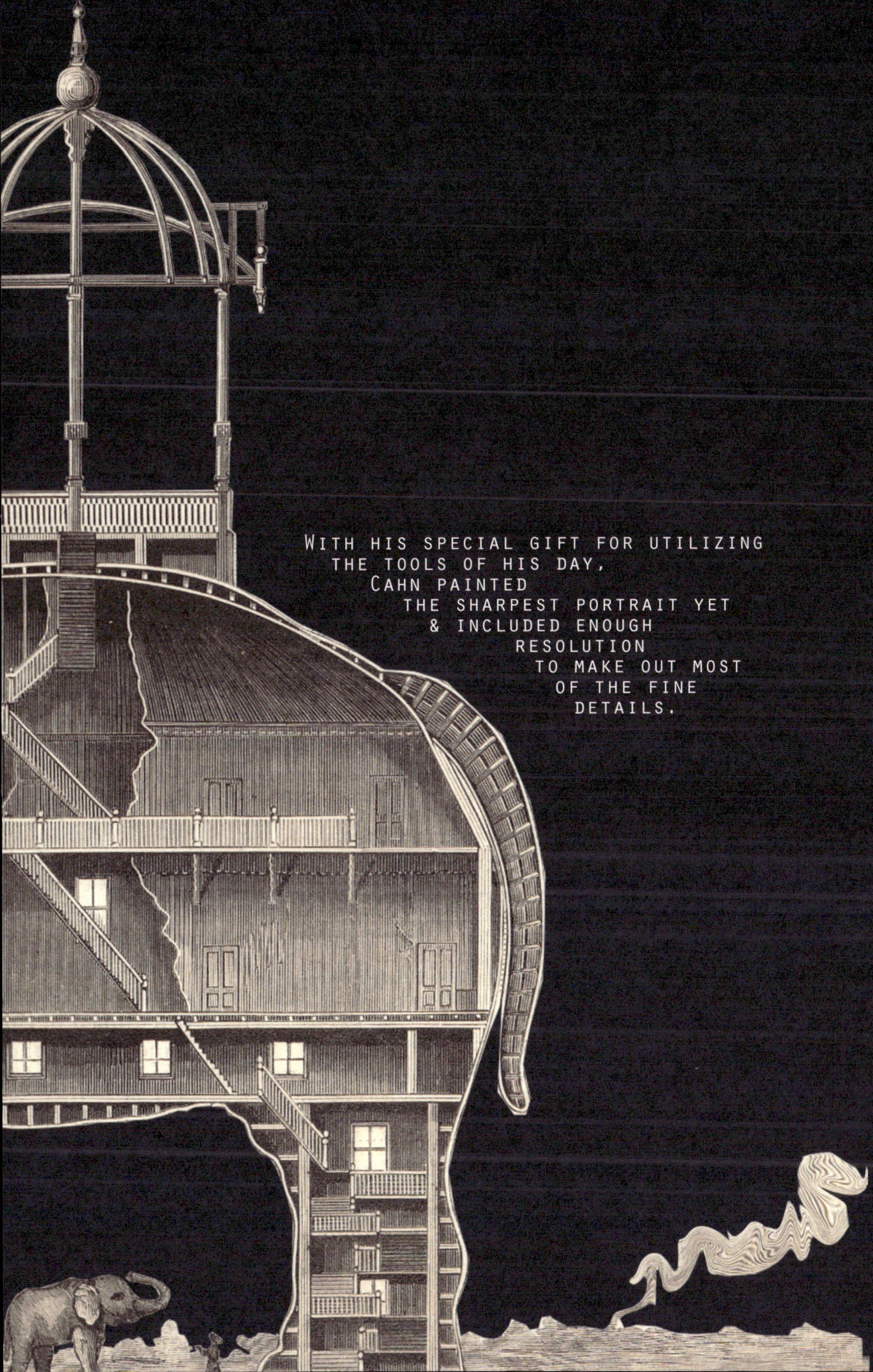

With his special gift for utilizing
the tools of his day,
Cahn painted
the sharpest portrait yet
& included enough
resolution
to make out most
of the fine
details.

Adams describes his colleague's discovery:

"Further work on [CBN's] chemistry was impeded by the fact that in spite of many attempts in different laboratories the isolation of cannabinol was not repeated until 1932. At that time Cahn, an English chemist, again obtained this compound & completed a series of brilliant researches from the results of which he was able to establish the skeleton & the substituents in the molecule...

but was unsuccessful in determining the orientation of all the groups... His evidence for this structure was conclusive except for the positions of the HYDROXYL & N-AMYL GROUPS."

Cahn's painstaking & inspired work used a method of extensive chemical degradations,

a burning release of the molecule's sidechains & liberation into their constituents.

He shook the elephant's skeleton apart with heat only to see if he could put the bones back together again.

He dropped the structures into his imagination's fiery swirl...

& tried to visualize how the devil you could assemble the pieces of this chemical crossword puzzle into a three-dimensional molecule.

CAHN FIGURED OUT ALMOST THE ENTIRE ELEMENTAL SKELETON OF CANNABINOL...

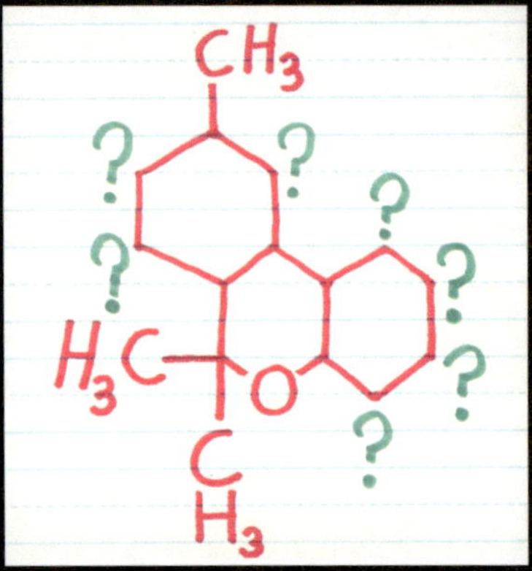

BUT STILL HE PUZZLED OVER WHERE TO...

PIN THE TAIL...

(n-amyl group)

& PLACE THE TOP HAT.

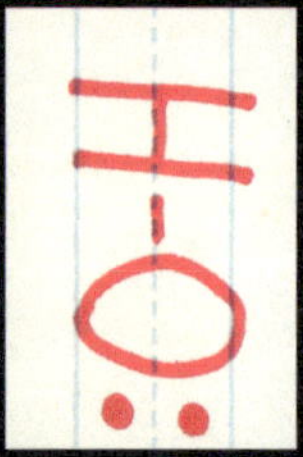

(hydroxyl group)

As for Old Bet, her story matched those of many immigrants flowing through the ports of New York,

their lives told by the Steinbecks, the Eisners & the muckraking lenses of Jacob Riis.

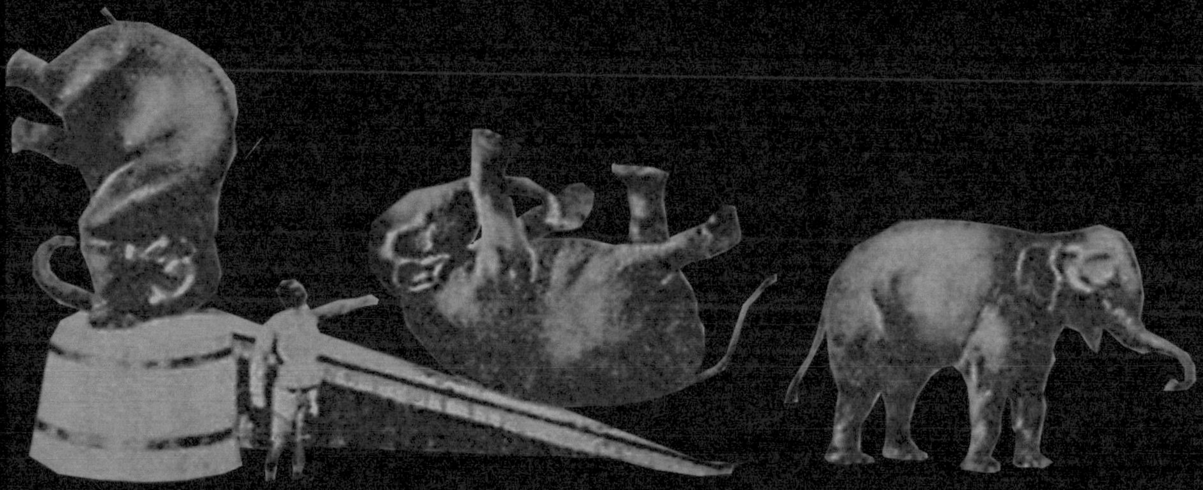

She worked the performance circuit through the South & Northeast,

transported by the grinding trains
to shows for ear-pulling yokels,

all while fed on a foreign diet of hay & oats

& in every town, sleeping in a different farmer's barn
(usually for a cut of the proceeds),

an elephant in chains prostituted for profit.

The cannabis plant knows all about hard use...

& advantages taken by a fair-weather friend.

During the American Revolution, hemp rope held the rebellious ships together & her canvas formed their sails.

Her fiber clothed their ragtag soldiers, formed the fuses of their furious Cornwall cannons...

& her hard-pressed paper caught the spark of ink & forever fixed the words of their feverish Declaration of Independence.

Hemp saw so many uses in the young nation that laws often required a certain acreage to always be planted by farmers. They needed prompting by these penalties because the processing of hemp's thick stalks into a usable fiber employed a process so brutal & dehumanizing that even Thomas Jefferson eventually relented, stopping his enslaved workers from retting & breaking it because it wore them down too rapidly.

All the way back to the very first agricultural books of China have growers been trading tricks for transforming their hemp into its useful husky Manila shine. Even as the invention of the cotton gin pushed hemp from the top shelf as the national clothing fabric, the plant continued its robust trade in the pharmacies almost up until World War II when again it was pressed into service for that fight. Hemp outfitted the US Navy in familiar patterns fixed into place ever since the triremes of the Peloponnesian War harnessed this fiber strong enough to resist the ever gnawing grind of the much salted sea.

But from the Sufis to the Taoists,
from the old witches of medieval Europe
& up until the flower children of today,

the presence of a psychoactive plant
often became a weapon against "Those People",

repeating a process so often racist, reactionary
& power-hungry that it's bluntly familiar
to anyone who sees the repetitious cycle
of the 'marihuana' of Mexico
evolve into the 'reefer' of the jazz era,
weaponize into the 'pot' of the longhairs
& finally transform into her most recent & fiendish manifestation
as a catch-all applied against all undesirables.

Who dares call the War on Drugs a failure?

As a piece of sociological technology,
it functions with utter humming perfection
as one of the finest pieces of autocratic bureaucracy
ever assembled in humanity's ever questing search
to stamp out the sins of their neighbors.

It stands to reason that Prohibition
– and its bastard child, the War on Drugs –
developed in this, the same nation who exported
to South Africa the system of apartheid whose laws
of patchwork horrors found its inspiration
in our own Jim Crow.

The grand finale of Old Bet's Amerikan Dream
was an ignominious end for us all.

A farmer wanted to fakt check
a rumor he heard that the thick hide
of an elephant could stop a bullet & so
hiding by the road for her walk
back to the barn after a show,
he fired his gun at her
& killed the pioneering pachyderm with
one shot right through the eye...

Or he did it because he loathed
the sinfulness of his fellow citizens
wasting their money on out-of-state
elephant entertainments...

Or he was driven mad by the loss of his
family grist mill.

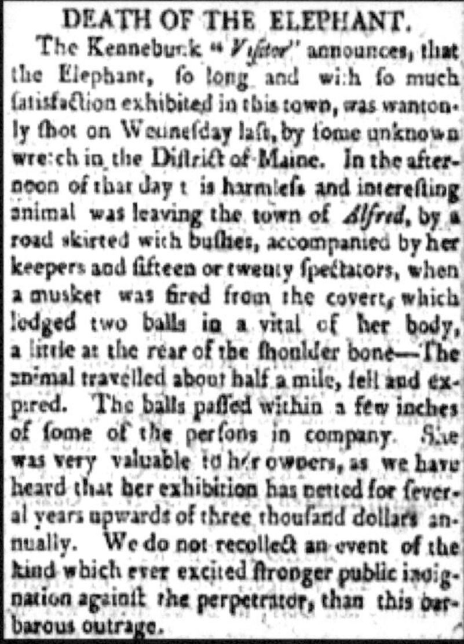

DEATH OF THE ELEPHANT.

The Kennebunk "*Vyſitor*" announces, that the Elephant, ſo long and with ſo much ſatisfaction exhibited in this town, was wantonly ſhot on Wednesday laſt, by ſome unknown wretch in the Diſtrict of Maine. In the afternoon of that day this harmleſs and intereſting animal was leaving the town of *Alfred*, by a road skirted with buſhes, accompanied by her keepers and fifteen or twenty ſpectators, when a musket was fired from the covert, which lodged two balls in a vital of her body, a little at the rear of the ſhoulder bone—The animal travelled about half a mile, fell and expired. The balls paſſed within a few inches of ſome of the perſons in company. She was very valuable to her owners, as we have heard that her exhibition has netted for ſeveral years upwards of three thouſand dollars annually. We do not recollect an event of the kind which ever excited ſtronger public indignation againſt the perpetrator, than this barbarous outrage.

So they say.

No one knows truly.

Rumors abound.

Other newspapers report other locations,
other circumstances;

unaligned stories by the lazy bastards
of the press muddy the waters.

And the greatest mystery of all
(if the story is even true):

how in heaven could that farmer
possibly murder an elephant
with a shot or two
from a farm gun?

The century following the death of Old Bet overflows with stories of elephants born in captivity or brought to the West in chains who went on to kill a person or four. For this crime & despite their immense value, they often received a death sentence. However, the executioners usually found those orders quite challenging to carry out:

It took over 100 rounds to kill the infamous Black Diamond Killer.

They filmed the electrocution of Topsy for posterity...

& the death marked an early union between Nikola Tesla's alternating current & the moving pictures of his rival, Thomas Edison.

In 1916, they hung murderous Mary.

Chunee's bloody execution in the London Zoo? 152 balls.

(& they still needed a harpoon to finish him)

So what's the biochemical equivalent of shooting an elephant?

For CBN, her monument stands in this flowchart from Dr. Roger Adams mapping out the last leg of the alchemical triangle:

SYNTHESIS OF THE MOLECULE FROM SCRATCH

Lock up your elephant in the basement & Frankenstein a clone out of parts.

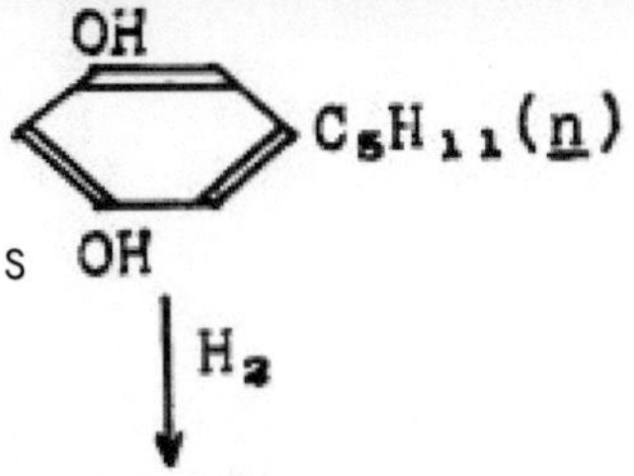

A FULL SYNTHESIS
FROM KNOWN
COMPONENTS...

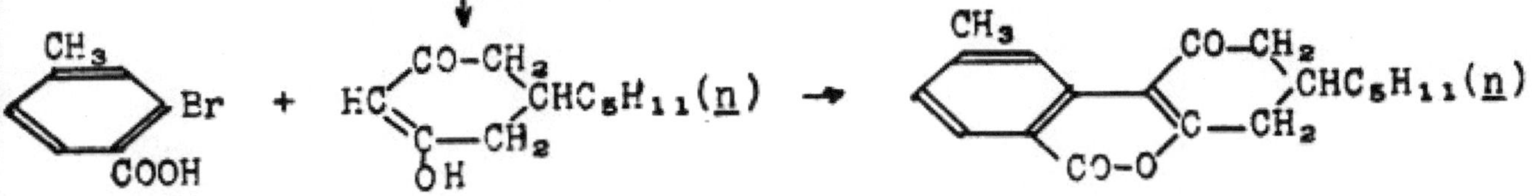

LAYS DOWN THE PATH TOWARDS A MOLECULE'S REALIZATION
& AFFIRMS OUR GUESSES ABOUT ITS NATURE.

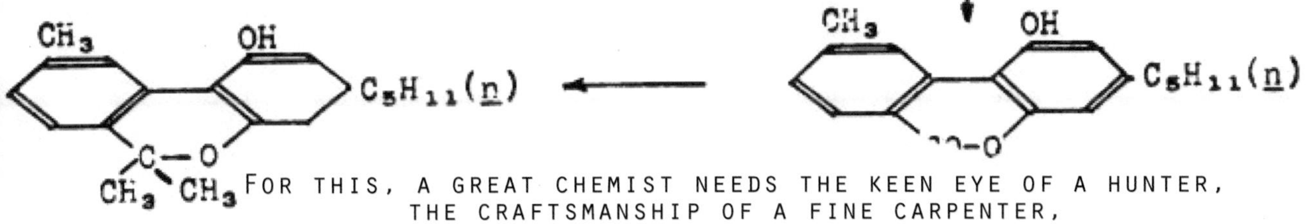

FOR THIS, A GREAT CHEMIST NEEDS THE KEEN EYE OF A HUNTER,
THE CRAFTSMANSHIP OF A FINE CARPENTER,
THE PATIENT ACCURACY OF A TAXIDERMIST,
& THE VISUALIZING IMAGINATIONS OF A GRAND ARCHITECT.

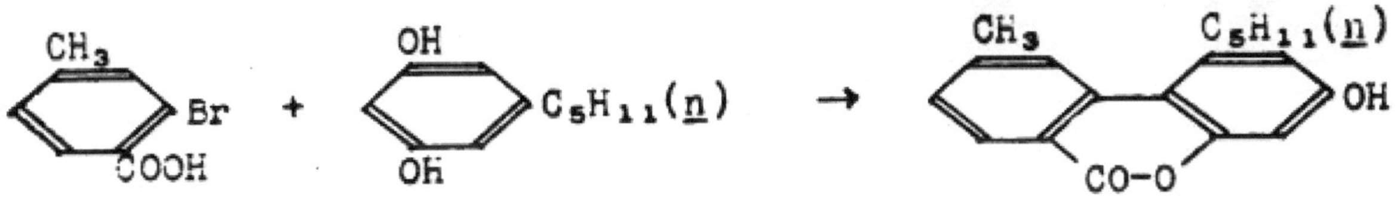

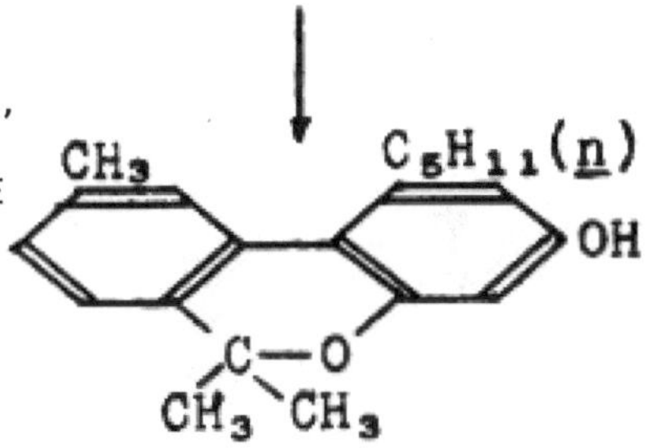

OF ALL THE "COMPLICATED & EXTENSIVE CHEMICAL INVESTIGATIONS" AROUND THE EARLY CANNABINOIDS, DR. ADAMS SAID THE WORK COULD ONLY BE PRESENTED IN "VERY BRIEF FORM & ONLY THE MORE SIGNIFICANT FACTS WILL BE MENTIONED"...

SO YOU KNOW THAT HERE YOU'RE GETTING ONLY A DRAUGHT,

– NAY, BUT THE DRAUGHT OF A DRAUGHT.
'OH, TIME, STRENGTH, CASH, & PATIENCE!'

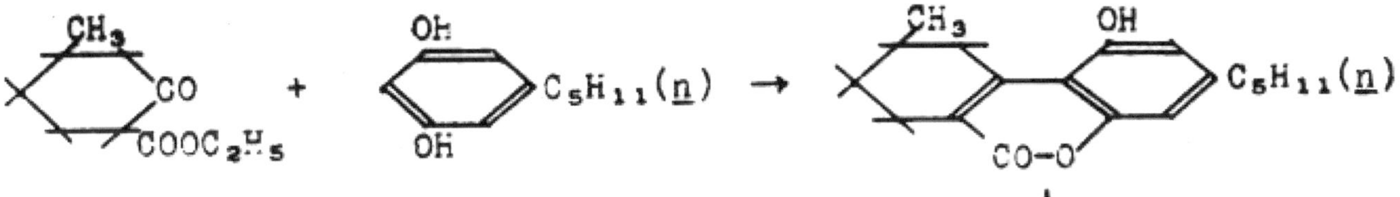

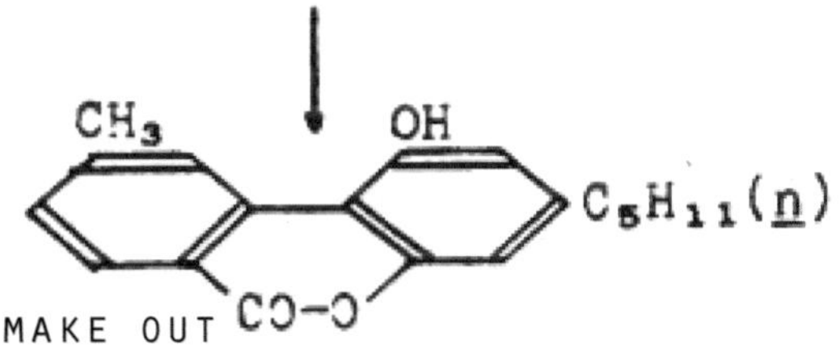

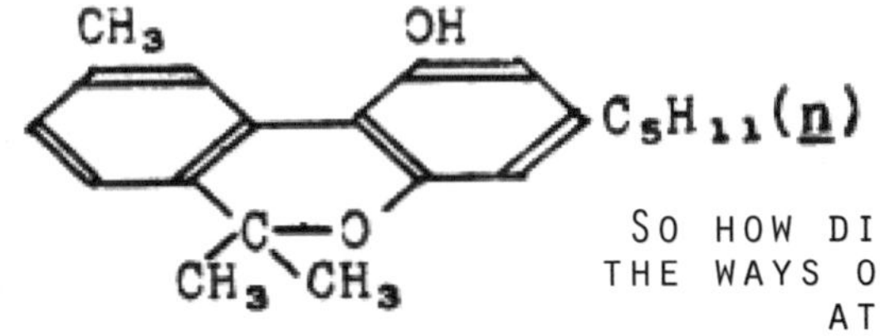

SO HOW DID ROGER ADAMS MAKE OUT THE WAYS OF CBN FROM THE GLIMMERS AT THE TOP OF THE TEST TUBE?

Adams needed creativity to place the last pieces & solve CBN's structure.

He started with a rainbow of color tests that indicated A PHENOL GROUP:

All the elegance of a six-membered aromatic ring wedded to a hydroxyl group (OH), an oxygen sharing hydrogen's lone electron.

And those double bonds on the inside of the well?

They stand for the three electron pairs shared between the six carbons of the ring, those particles of negative charge always whipping around trying to keep many masters happy.

In addition to the phenols, from his chemical hatchet jobs on CBN, Adams found that the molecules emerging from the solution often tended to be...

A good detective like Adams would see all those oxygens popping up in the middle & suspect two phenol rings...

as indeed it has.

In the midst of these investigations, Adams also worked to crack CBD.

While working on the problem, he realized that when you ripped CBD in half, you often found an OLIVETOL.

This final clue gave him the secret as to the location of CBN's hydroxyl & n-amyl groups – the missing top hat & tail.

The linkage joining those two benzene rings must lay between the hydroxyl (OH) groups.

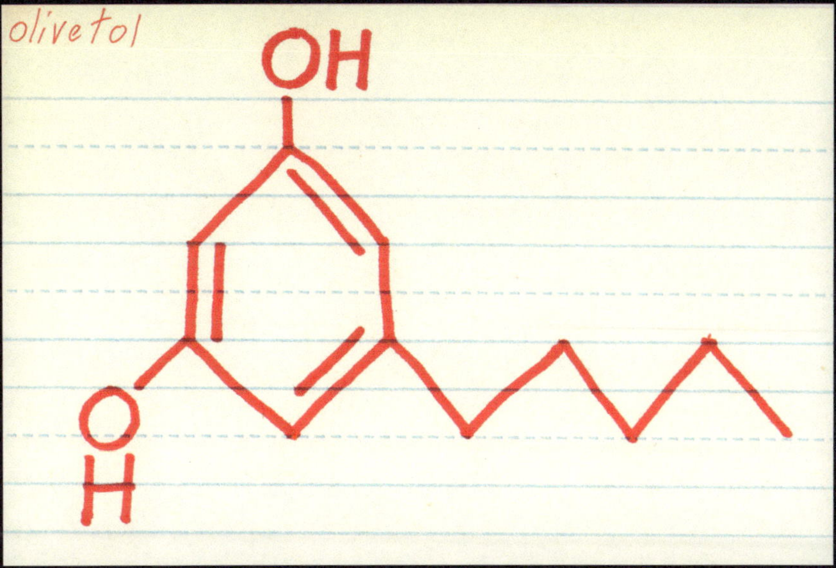

Once he knew what CBN looked like, Adams proved his vision correct by performing its synthesis with the pathway he pioneered.

In 1940, as war waged in Europe, Adams ended a pursuit that had ran over 80 years: the search for the structure of a cannabinoid.

THUS THIS VOLUME CLOSES
WITH TWO IMPRESSIVE SKELETONS...

NOTICE.

THE Skeleton of that unfortunate Elephant that was shot the 26th of July last, in District of Maine, so well known by the public, is got up for inspection and may be seen at No. 301 Broadway, from Monday the 7th until Wednesday the 30th instant, every day in the week, Sunday excepted, from nine in the morning until sunset—The weight of the Elephant, when it was shot, was upwards of 70 hundred pounds.

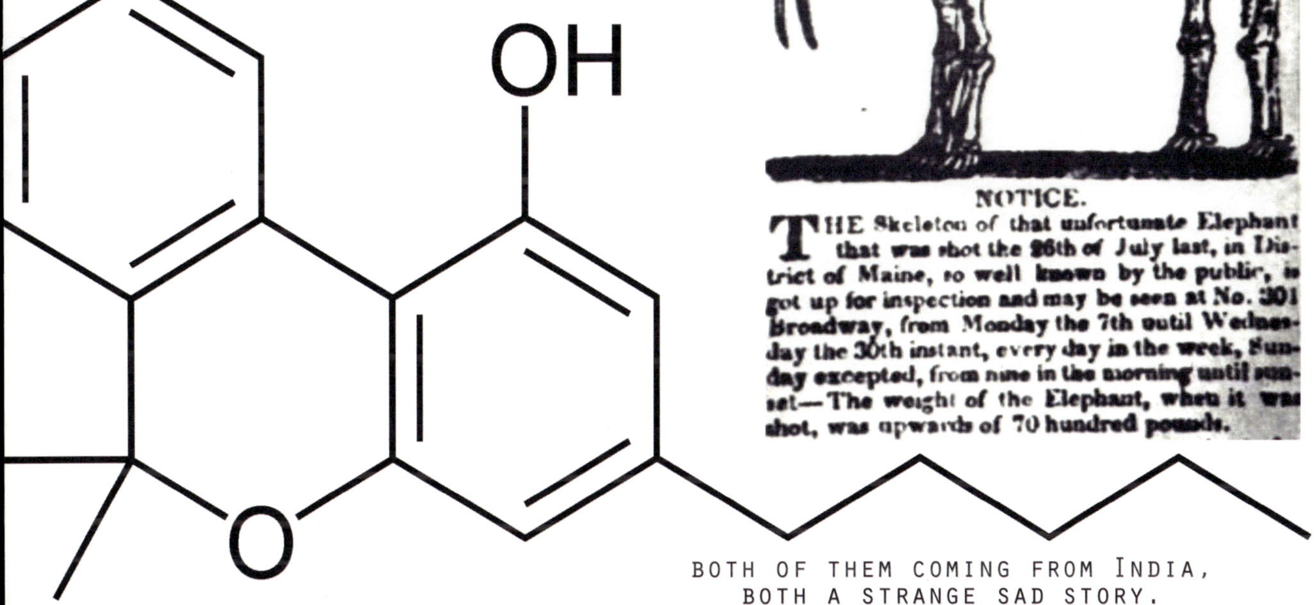

BOTH OF THEM COMING FROM INDIA,
BOTH A STRANGE SAD STORY.

In his Harvey lecture of 1942, Adams ends with a tantalizing hint about the team of Wollner, Matchett, Levine & Loewe finding great activity from an isolated isomeric tetrahydrocannabinol.

We must also never forget to pay our respects
to the memories of those souls lost while diving
for the Test Tube's translucent bottom.

In Roald Hoffman's "The Same & Not the Same",
the Nobel Laureate's classic collection of essays
on the shrouded mysteries of chemistry
& the complex emotions of its seekers,

the doctor states,
"chemists are no more reflective
than other people. But the questions they pose
& the craftsmanship with which they answer them,
move them to consider polarities
& the associated tensions" while they
try to figure out their
unknown samples.

It leads them "to reflect on whether the substance
made is the same or not the same as others...
Why is that question interesting?"

"Because the question of identity,
of our identity, shaped in childhood
in a complex dance of bonding & separation,
matters deeply to us. The processes of nature
connect with the interior world of our emotions."

"Identity & deception, origins, good & evil,
sharing & withholding, resurrection,
danger & safety & overcoming obstacles
are some of the psychological constructs
or mystical structures with which the world of
molecules connects. These emotional focal points
shape, consciously & subconsciously,
the wonderful, gameplaying psychology
of the chemist engrossed in molecules. It helps
us to see this, to sense what moves chemists
& I think the material-psychological link,
expressed through polarities, allows us
to understand why we like
& fear chemicals."

CHEMISTS SEEKING TO WALK WITH LADY TRUTH NEED PERSISTENCE
BECAUSE SHE SO OFTEN GOES BY HER OWN WAY ALONE,
ALWAYS LEAVING THE QUESTION:
HOW TO MAKE IT ALL VISIBLE?

WHAT EYES CAN SEE FAR BELOW THE WAVES OF LIGHT
WHERE ELECTRONS CAVORT LIKE CHARGED DOLPHINS,
UNCERTAIN AS HEISENBERG ON WHERE
THEY'LL JUMP NEXT?

HOW TO TRANQUILIZE A MOLECULE SO YOU CAN MEASURE ITS HIDE,
TRACE THE BONES UNDER THE HYDROGEN SKIN,
& PEER AT ITS BENZENE RINGS TO SEE HOW THEY SPIN?

HOW TO WONDER AT ITS SPIRAL OF LIFE:
WHERE DOES IT SPEND ITS TIME?
WHO DOES IT LIKE & WHO DOES IT LOATHE?
HOW DO THEY WOO ONE ANOTHER?
AND IS SYNTHESIS AS SEXY AS SEX?

VISIONARY ADVOCATES LIKE DR. HOFFMAN GAZE DOWN SIMPLY
INTO THE TEST TUBE'S BOTTOM JUST TO FIND ANOTHER MIRROR.

FIN

THAT'S ALL FOLKS!

SEE YOU NEXT TIME
WHEN WE CONTINUE
THE CHASE IN...

VOLUME 2: ELEPHANTIZING,
THE JUMBO STORY OF THC

TURTLES ALL THE WAY DOWN...

SOURCES, IMAGES, DIGRESSIONS & FAKTS

The spelling of Amerika comes from Kafka's novel of the same name, a place he never visited. Fakt comes from the animation 'Harvie Krumpet', an inspiration to us all.

Thomas Nast's election cartoon of 1877

La Morphiniste by Eugène Grasset (1897)

In the Arms of Morpheus by W. Reynolds-Stephens(1894)

The Demon Behemoth by Louis Le Breton in Collin de Plancy's 'Dictionnaire infernal' (1863)

from Doré's Dante: evil counselors

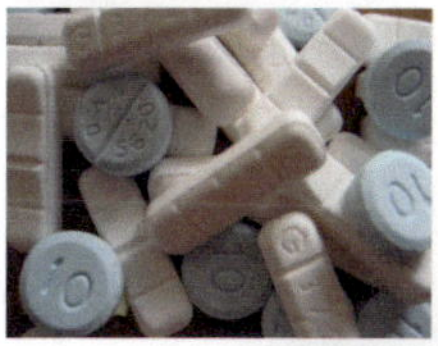

Source: Dean812 on Flickr

Katherine Hepburn in Eugene O'Neil's 'Long Day's Journey Into Night'

Phoenix from 'The Hieroglyphics of Horapollo

'DRUG ABUSE' FUZZY WITH DISAGREEMENTS

In the mid-60s, the docs, the cops & the eggheads all held different meanings of the term 'drug abuse'. The report of Prettyman to JFK & LBJ offered a rather standard definition: "Drug abuse is frequently considered to mean only narcotic addiction classically, however, addiction involves:

1. Physical dependence with a withdrawal state
2. The building of tolerance to the same dose
3. Psychological dependence on the drug, characterized by an overpowering desire or compulsion to continue taking it."

But his commission impressed again as being ahead of the curve with their notes on the 'spree' style of drug taking: "On the other hand, in many instances of drug abuse there is little true addiction. In the 'spree' use of drugs (the taking of different drugs in rotation), there is, for example, usually no development of physical dependence or of tolerance. Similarly, certain drugs, such as a stimulant & those which produce distortions in perception [psychedelics], may not produce physical dependence or tolerance even when taken continuously. Yet in all these instances the drug abuser can harm both himself & society. While the specific problem of narcotic addiction is serious & difficult, it is only one part of the much larger problem of drug abuse."

Digging into the report, they showcase the clear & simple truth: the vast majority of Americans who consume drugs suffer no serious long term consequences from doing so.

THE END OF CAMELOT

A book to read is the 'Yankee & Cowboy War'. The heavy to watch is Allen Dulles in the 'The Devil's Chessboard'.

Howard Pyle's 1902 "The Fishing of Thor and Hymir" North Folk Legends of the Sea Recovered

Gustave Doré's 'Gargantua & Pantagruel' by Rabelais (the greatest cannabis author of the Renaissance as we shall someday see...)

caption: The German astroler Trippa predicts that if Panurge marries he will be cuckolded (III, 25)

Albrecht Dürer's 'On the Art of Measuring. A Man Drawing a Lute'

& his spot-on illustrations for Brant's 'Ship of Fools'

& Doré's 'A History of Holy Russia'

FATHER OF THE CANNABINOIDS: Mechoulam, Gaoni & Other Idols

In science journalism, it's far too easy to fall into the trap of laying almost every modern discovery at just one pair of feet or to one team of thinkers who shared the Nobel Prize for a breakthrough. But all in the trenches of real science know these Big Names always rest on a pyramid of grad students, a global web of collaborators, a pool of research techs & postdocs working alongside the scientists of industry, the feds writing the checks & friendly labs next door loaning their biochemical tools & talent.

Dr. Mechoulam has always been quick with giving credit to his collaborators & I've heard a number of his co–workers cite his inclusive leadership as instrumental in facilitating the explosion in cannabinoid chemistry since 1964. For those many other names on the papers of the peer–reviewed literature, I aim to cover as many of them as I can in this long meandering voyage but for now I will stick to the slang of the Big Man narrative hypothesis just for the sake of this trumpeting chapter.

For your best chance to scope out the many researchers behind the multilayered discoveries of the cannabinoid systems, take an hour to watch the free documentary 'The Scientist'. Then go to a conference to meet the scientists in the flesh. They're all very approachable. Even to young drug writers without yet a book to their name.

undated illustration of a meteorologist working at a desk in the "Old Probabilities" instrument room of the Signal Service Bureau in Washington, DC

Heinrich Khunrath's 'The Oratory & the Laboratory' with its 3d 7-leaved chandelier...

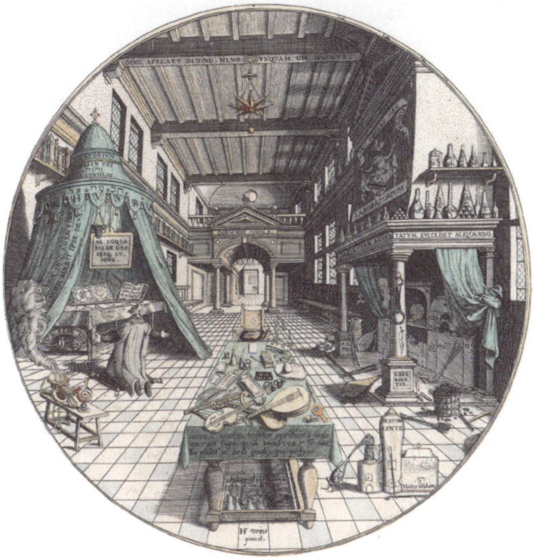

under the inscription 'Without the breath of inspiration from God, no one finds the great way."
(see Christ Bennett's essay 'Cannabis The Herb of Saturn' or his detailed canna-alchemy book 'Liber 420')

THE BEAM TEST

Once the cat was mostly out of the bag about the unreliability of the Beam test to detect cannabinoids, you should have seen the backbending at the great Marihuana Conference of '38 by the Federal Bureau of Narcotics chemist Dr. Levine. He readily admits the Beam test fails in some cases but gives a whole host of alchemical tricks to make it work better, sounding like a used car salesman trying to repaint a pink elephant. The transcript can be found in the online archives of the Schaffer Library of Drug Policy, a treasure trove of history that yields constant reminders that there's nothing new in the federal stubbornness to stand by lousy science: http://www.druglibrary.org/schaffer/hemp/taxact/mhc11.htm

from Salonmon's Trismosin's 'Splendor Solis'

Advertising cards "Wear Celluloid collars and cuffs" from Boston Public Library

THE HARE & THE FROG

Hare's frog experiments also stand as a powerful piece of data presaging the number one reason given at every dispensary for why people are using medical cannabis: pain.

Hare stated, "I have found the efficient dose of a pure extract of hemp to be as powerful in relieving pain as a corresponding dose of the same preparation of opium" & declared the advantage to cannabis because it left one less soporific & rarely induced the nausea more common to opiate–based medicine.

When the good Dr. Hare described the painkilling effects in 1887, he wrote, "During the time that this remarkable drug is relieving pain a very curious physical condition sometimes manifests itself; namely, that the dim intuition of the pain seems to be due to its fading away in the distance, so that the pain in a delicate year would grow less & less as a beaten drum was carried farther & farther out of the range of hearing."

HARE'S HILARITY & DENTISTRY

As a fair physician, Dr. Hare self–administered the medicine he recommended to his patients: "I myself have taken as much as one grain of the solid extract of a very active preparation without producing any disagreeable symptoms other than that of a deep sleep, which lasted for nearly 18 hours, preceded by a period of great hilarity, which did not pass into any sensation of dread, such as has been described by some persons."

(OH JASPER, DOES THAT DESCRIPTION NOT MAKE YE WISH TO FIRE UP THE TIME TWISTER MACHINE & GO BE A FLY ON THE WALL OF THE GOOD DOCTOR HOBART HARE'S LAB, WATCHING FOR YOURSELF THE GREAT HILARITIES?)

With two intriguing hints Dr. Hare ends his monograph from just after the Civil War: "I desire to call the attention of practitioners to the local anesthetic action of this curious substance. As already detailed, when applied directly to the nerve–trunks it paralyzes them, & I have found that when applied to the mucous membranes of the tongue in considerable quantity it diminished sensibility to a considerable degree. Dentists, I am told, constantly use it for sensitive dentine."

(BY THE BULL OF SHIVA! INGESTING POT FOR THE PAIN OF PULLING A TOOTH? THEN THAT OLD TROPE MIGHT BE TRUE? HOW MANY TENS OF THOUSANDS OF STONERS MUST HAVE FANTASIZED ABOUT BEING ALLOWED TO BURN ONE DOWN IN THE WAITING ROOM IN ORDER TO STEADY THE NERVES & STRENGTHEN THE SINEWS?)

Of course, the irony on pain & cannabis is that if anything's proven by modern science, it's that this plant is biphasic. In many aspects of its medical efficacy, a little bit usually helps – but a lot often hinders. In general, tests have found that when wicked stoned, painful stimuli hurts worse. So to end on the highest note we can while still invoking dentistry, imagine dear friends, a brave new world where your oral hygienist smears a cannabis salve on your gums to lessen sensation & offers a few hits off a high CBD vape pen before the procedure.

Beloved Cruikshank's 'The Head Ache'

from John Flaxman's Odyssey

PLUTO'S TALE
(a Sumaq story, as told to Lex)

Pluto was a wonderful dog. He loved hanging out with people & he adored Sumaq's father. But after 18 years, he had all kinds of pain. Especially from a chronic prostate problem that troubled him greatly. Because of his age, they couldn't give him any painkillers.

On Pluto's last night on earth, Sumaq woke up in the middle of the night to realize Pluto was near the end. He whimpered in great pain & nothing interested him. So she gave her old friend the medicine. When she took a hit of weed & shared the smoke to him, it was the first time Sumaq's father ever saw her smoking pot. But when he saw the quick relief for his dog, he was convinced. The cannabis gave Pluto back his appetite & before they took him to the vet for the long sleep, he ate an entire juicy delicious hamburger for his last meal.

Praise be to the medicine that helps us all creatures under the sun.

from Doré's Fables

Planentary Systems, Woodcut from an edition of Boethius Consolation published at Augsburg 1537

Léon Benett's illustration of Jules Verne's mechanical elephant from 'The Steam House' (1880)

An elephant keeper riding his elephant from 'Tashrih al-aqvam' (1825)

Joseph Nash's illustration of Prince Albert Victor in India, Elephant catching in Mysore (1890)

THE

ELEPHANT,

1797 Apr 23 - broadside for exhibition at Beaver & Broadway

THE ELEPHANT & ARISTOTLE

The essay 'Aristotle's Notices on the Elephant' by Martin Peris is a loving piece about the First Biologist & his keen eye for the ins & outs of life's largest quadruped. However, Peris does mention that the gift of an elephant from Alexander the Great might have actually been a mean practical joke, if his his former student gave him the beast but not the money to feed it.

In any case, Aristotle appears to have extensive firsthand experience with them ("elephants copulate in lonely places"). He marvels at the intricacies of their trunk, a prehensile appendage used with the same multivariate dexterity of our own much celebrated hand. Perhaps via a dissection did he realize that the elephant's trunk contains no bones, just many many muscles. According to Peris, Aristotle didn't hazard a guess to the number of muscles, but "as is well known to elephantologists", the celebrated Cuvier estimated 40,000 muscles in an elephant's trunk – an estimate that turned out to be low by a factor of ~4.

The Haschish

Of all that Orient lands can vaunt
Of marvels with our own competing,
The strangest is the Haschish plant,
And what will follow on its eating.

What pictures to the taster rise,
Of Dervish or of Almeh dances!
Of Eblis, or of Paradise,
Set all aglow with Houri glances!

The poppy visions of Cathay,
The heavy beer-trance of the Suabian;
The wizard lights and demon play
Of nights Walpurgis and Arabian!

The Mollah and the Christian dog
Change place in mad metempsychosis;
The Muezzin climbs the synagogue,
The Rabbi shakes his beard at Moses!

The Arab by his desert well
Sits choosing from some Caliph's daughters,
And hears his single camel's bell
Sound welcome to his regal quarters.

The Koran's reader makes complaint
Of Shitan dancing on and off it;
The robber offers alms, the saint
Drinks Tokay and blasphemes the Prophet.

Such scenes that Eastern plant awakes;
But we have one ordained to beat it,
The Haschish of the West, which makes
Or fools or knaves of all who eat it.

The preacher eats, and straight appears
His Bible in a new translation;
Its angels negro overseers,
And Heaven itself a snug plantation!

The man of peace, about whose dreams
The sweet millennial angels cluster,
Takes the mad weed, and plots and schemes,
A raving Cuban filibuster!

The noisiest Democrat, with ease,
It turns to Slavery's parish beadle;
The shrewdest statesman eats and sees
Due southward point the polar needle.

The Judge partakes, and sits erelong
Upon his bench a railing blackguard;
Decides off-hand that right is wrong,
And reads the ten commandments backward.

O potent plant! so rare a taste
Has never Turk or Gentoo gotten;
The hempen Haschish of the East
Is powerless to our Western Cotton!

- John Greenleaf Whittier (1858)

(THIS QUAKER ABOLITIONIST ALSO WINS THE AWARD FOR 'MOST APPROPRIATE NAME FOR THE FIRST POET TO WRITE ABOUT CANNABIS IN THE US')

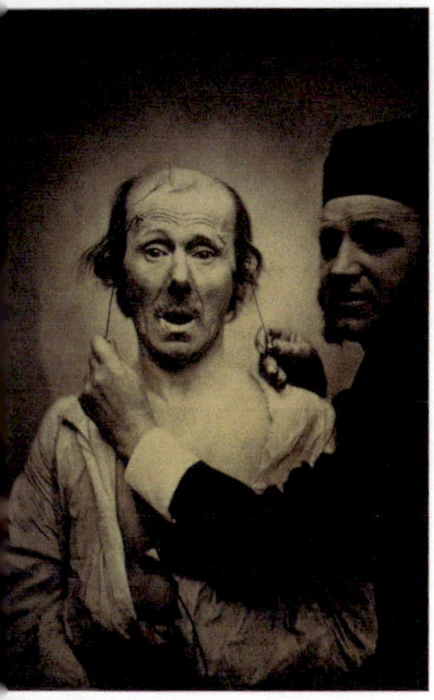

DARWIN, DUCHENNE & THE OLD MAN

These photos appear in Darwin's 'The Expression of the Emotions in Man & Animals', one of his most intimate books & one of the earliest scientific works to use evidence from the still novel technology of photography.

Here his collaborator Duchenne de Boulogne, a French neurologist, shows a tool used in his studies of muscles & electricity. The good doctor would insert sharp needles underneath the skin of human test subjects & then with an applied jolt of juice from his appareil volta–électrique, Duchenne revived the early work of Galvani from 70 years prior & used the electricity to stimulate various combinations of the 43 facial muscles, searching for the patterns that produced our most common emotions. He identified 13 primary emotions controlled by one or two muscles.

Unsurprisingly, critics attacked both scientists because this obviously painful procedure was carried out on patients like this "old toothless man" who the Frenchman described as possessing "a thin face, whose features, without being absolutely ugly, approached ordinary triviality."

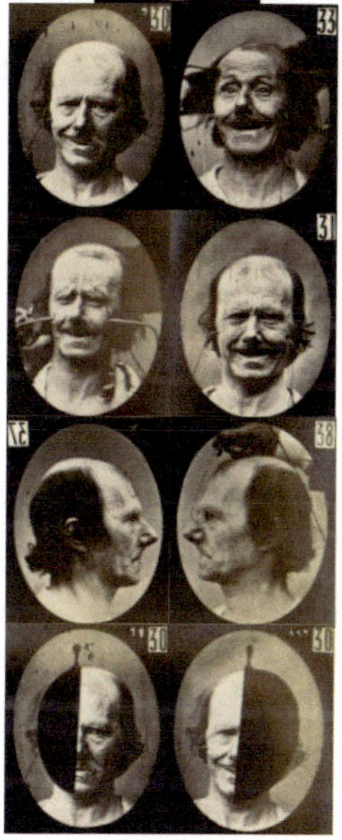

However, Duchenne argued that an anesthetic skin condition made the old man's face tough as leather & barely registering the shocks. Furthermore, the man said he didn't mind them taking photographs of his electrically grimaced face in their pursuit of the muscles underlying emotional expressions, the "gymnastics of the soul".

Duchenne argued democratically that "every face could become spiritually beautiful through the accurate rendering of his or her emotions." Darwin used these images in his pioneering study of emotions & it consistently amazed him that when friends were shown the old man's photographs, they usually guessed correctly at the apparent emotion induced by the electrical shocks.

Darwin even invoked the Bard to describe the old man's face:
Thus Shakespeare says, "I saw a smith stand with open mouth swallowing a tailor's news."
('King John,' act iv. scene ii.)

ENSLAVEMENT & ELEPHANTS

We'll use this image of an elephant because it comes from an 1887 report to King Leopold II & nothing symbolizes the viciousness of colonial rule more than that blackhearted monarch who ran the Congo as a personal fiefdom for converting rubber & slave labor into piles of cash for beautifying his Belgium. The book to read is Hochschild's 'King Leopold's Ghost: A Story of Greed, Terror, & Heroism in Colonial Africa'.

the infamous diagram of the stowage of enslaved African people on a British ship

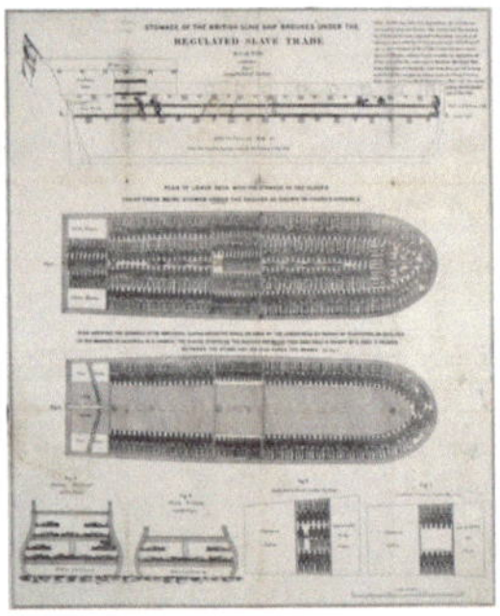

an ad for Moriz beer & Guiness beer

Two Bit for Old Bet

Perhaps the elephant viewing fee of a quarter or fifty cents inspired this jump–rope rhyme of the day that's been called "the most common hand–clapping game in the English–speaking world":

Miss Mary Mack, Mack, Mack
All dressed in black, black, black
With silver buttons, buttons, buttons [butt'ns]
All down her back, back, back.

She could not read, read, read
She could not write, write, write,
But she could smoke, smoke, smoke
Her father's, pipe, pipe, pipe)

She asked her mother, mother, mother
for fifty cents, cents, cents
To see the elephants, elephants, elephants
Jump the fence, fence, fence.

They jumped so high, high, high
they reached the sky, sky, sky
And didn't come back, back, back
Till the 4th of July, 'ly, 'ly!

July can't walk, walk, walk
July can't talk, talk, talk
July can't eat, eat, eat
With a knife and fork, fork, fork.

She asked her mother, mother, mother
For 5 cents more, more, more
To see the elephants, elephants, elephants
Jump over the door, door, door.

They jumped so low, low, low
They stubbed their toe, toe, toe
And that was the end, end, end,
Of the elephant show, show, show!

(perhaps parallel to cannabis inspiring the ancient Taoist poetry penned to Ma Gu, aka Miss Hemp)

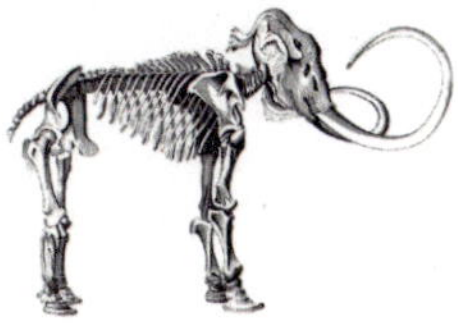

Illustration of a Mammuthus primigenius (1880s)

Cave painting in the Grotte de Rouffignac

Justus van Bentum's 'The Explosion in the Alchemist's Laboratory' (~1700)

Rockwell Kent's 'Moby Dick'

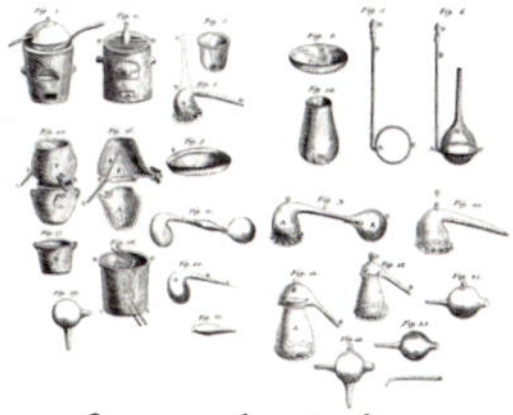

from Antoine Laurent's 'The Elements of Chemistry'

from Alchemical and Rosicrucian Compendium (~1780 in the lower Rhine)

from 'The Twelve Keys of Basil Valentine'

19th-century engraving of Goethe's Faust and Homunculus

from 'Donum Dei' (17th century)

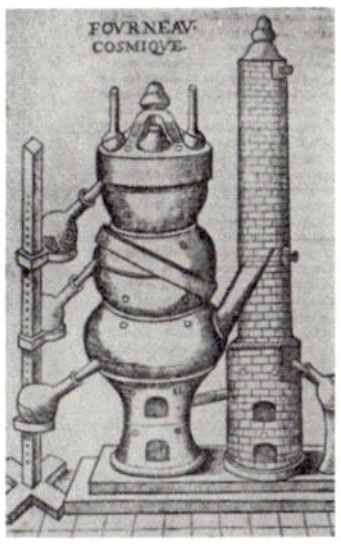

The Cosmic Furnace from Annibal Barlet's 'Le Vray Cours de Physique' (1653)

from Geber's Liber Fornacum

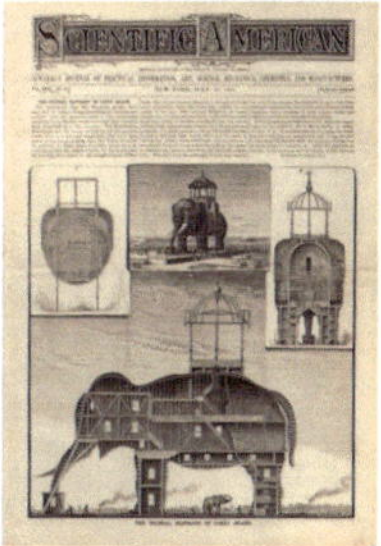

The Elephantine Colossus in Coney Island (Scientific American)

Howard Pyle's Lady Truth

THE CBN SYNTHESIS OF ADAMS

2-Bromo-4-methylbenzoic acid & dihydroolivetol condensed in the presence of alkali & a copper salt to i-keto-3-n-amyl-9-methyl-1,2, 3,4-tetrahydro-6-dibenzopyrone; dehydrogenation gave i-hydroxy-3-n-amyl-g-methyl-6-dibenzopyrone which, with excess of methylmagnesium iodide, gave cannabinol, (I-hydroxy-3-n-amyl-6,6,9-trimethyl-6-dibenzopyran). An analogous condensation of 2-bromo-4-methylbenzoic acid with olivetol, followed by treatment with methylmagnesium iodide, gave the isomeric cannabinol with the linkage between an hydroxyl & the n-amyl group. The second method consisted in condensation of ethyl 5-methylcyclohexanone-2-carboxylate with olivetol in the presence of phosphorous oxychloride to give i-hydroxy-3-n-amyl-9-methyl-7,8,9, i o-tetrahydro-6-dibenzopyrone. Dehydrogenation to the corresponding I-hydroxy-3-n-amyl-g-methyl-6-dibenzopyrone followed by treatment with methylmagnesium iodide gave cannabinol.